the ultimate AÇAI SMOOTHIE *cookbook*

MORE THAN 120 SMOOTHIE RECIPES
MADE WITH THE AGE-DEFYING AÇAI BERRY

also by
DONNA J PLINER RODNITZKY

Ultimate Smoothies
Summer Smoothies
Slim Smoothies
Low-Carb Smoothies
Sinful Smoothies
Tipsy Smoothies
Ultimate Juicing
The Ultimate Low-Carb Diet Cookbook
The Ultimate Cold Slab Ice Cream Cookbook

the ultimate AÇAI
SMOOTHIE *cookbook*

MORE THAN 120 SMOOTHIE RECIPES
MADE WITH THE AGE-DEFYING AÇAI BERRY

by

DONNA J PLINER
RODNITZKY

Printed in Canada

Design by Wendy Norwood

ISBN 978-1-59975-962-3

First Edition

*I am grateful to my husband Bob,
who encouraged me to write
The Ultimate Açai Smoothie Cookbook,
and critiqued the recipes.*

*I also would like to thank my daughter Laura,
a gifted writer, who both edited and
contributed to the book.*

*Finally, I would like to acknowledge
Wendy Norwood, who designed the book.
She has been a wonderful resource in
bringing this book to fruition.*

introduction

introduction

❦

❝ *The greatest wealth is health.* **❞**

—**Virgil**

Spanish explorer Ponce de León famously failed in his search for the Fountain of Youth, but, actually, he wasn't that far off. Had León just gone a bit further south, he might have come across a secret that had been kept for centuries by the indigenous tribes and medicine doctors of the Amazon and other parts of South America. The remarkable açai (ah-sigh-ee) berry, found growing wild in this region, possesses healing powers previously unknown to the Western world. The indigenous peoples of the Amazon rainforest discovered long ago that this miraculous berry enhanced their health, stamina, and vitality. Scientists today believe the rainforests of the world may hold the keys to treating and preventing many of our most feared diseases.

Among the more than 200 fruits found in the Amazon, the açai berry is considered the most nutritive – in fact, some experts believe açai is the single most nutritive food in the world. This small "superberry" is packed with vitamins, minerals, and antioxidants that provide an endless list of physical and mental benefits to people of all ages. It may indeed be the closest we'll ever get to finding the fountain of youth.

The açai berry is 100% natural and has just about everything you need to maintain good health: Vitamin B1, Vitamin B2, Vitamin B3, Vitamin E, Vitamin C, phosphorus, calcium, potassium, iron, fiber, proteins, and very high amounts of essential fatty acids and Omega-6 and Omega-9. Moreover, it's low in sugar and is believed to have up to 33 times the antioxidant value as red wine grapes.

While many of the health claims of açai are still being investigated, scientists have already confirmed several of the amazing benefits this little berry has on the human body and mind. Here are just a few:

• **Mental and Heart Health:** The açai berry cleans and detoxifies the body, improves blood circulation (by improving the health of the heart and the cardiovascular system with its high level of antioxidants), enhances sleep, and boosts the immune system, all of which can lead to improved mental activity and overall vitality and stamina.

• **Lower Cholesterol Levels:** Açai berries are known to contain large amounts of protein and fiber, which may help lower cholesterol. Moreover, they contain Omega-6, which lowers cholesterol levels, and Omega-9 fatty acids, which lower LDL cholesterol levels and maintain natural HDL cholesterol levels. It is also believed that these omega fatty acids protect the heart from damage and reduce the amount of cholesterol in the blood, thereby providing possible protection from heart attacks, strokes, and other common cardiovascular complications. They also aid in the absorption of certain vitamins, such as Vitamins A, D, E, and K.

• **Anti-Aging:** Many people are enthusiastic about the açai berry's potential anti-aging properties. Açai is rich in amino acids and vital trace minerals that aid in proper muscle contraction and regeneration, thereby slowing down many of the signs of aging. The berry is also high in monounsaturated oleic acid: This fatty acid combined with Omega-3 fish oils penetrates cell membranes and makes them suppler, allowing hormones, neurotransmitters, and insulin receptors to operate more efficiently. This, in turn, greatly reduces inflammation, which is one of the leading causes of aging.

• **High in Antioxidants:** The antioxidant benefits of the açai berry, combined with its fatty acids, make açai one of the most important superfoods on the market. The açai berry is believed to have an antioxidant content up to 33 times greater than the antioxidant content of red wine grapes and twice as high as that found in blueberries. In fact, based on this amazing property, the açai berry may potentially be used to treat a number of medical disorders, such as cancer.

• **Increased Energy and Stamina:** Açai is known to give you an energy boost. Many people report feeling younger and more vital shortly after adding açai berry to their daily diet, with a positive emotional lift and an overall feeling of well- being. What's more, both men and women report an increased improvement in their sexual function and/or libido. It's no wonder, then, that açai is known as "Amazon's Viagra".

We can thank the California-based company SAMBA, Inc. (Saving and Managing the Brazilian Amazon) for bringing açai out of the rainforests and into our homes. SAMBA's initial mission – to protect the diversity of the rainforest and improve the socio-economic conditions of the local people – led the company to work with impoverished fruit collectors throughout the lush floodplains of the Amazon River in Northern Brazil. It was during this effort that the founders had the good fortune to discover the açai palmberry, which grows on top of palm trees (euterpe oleracea) found only in the Amazon jungle. The small, round, blackish açai berries hang from the branches in clusters and makes quite an impressive picture.

Before you think, "Of course, they went to help out and instead turned it into a commercial business," consider this: The açai fruit can only be picked by hand, so the berries are collected by local families. The açai boom is thus providing employment for area workers and is proving to be a successful alternative to the logging and forestry industries that have led to so much destruction in the rainforests.

The Fountain of Youth? The world's most perfect food? The best-tasting medicine you've ever had? Or just a simple berry that adds a fruity and chocolatey flavor to your daily smoothie? There's only one way to find out. In this book you'll find more than 120 ways to enjoy açai. Try a smoothie a day – or two or three – and experience for yourself the energy-boosting and overall health benefits of açai.

Smoothie Fruit — Açai Supporting Cast

HOW TO SELECT, PREPARE, AND STORE FRESH FRUIT

❦

Time flies like an arrow; fruit flies like a banana.
—**Groucho Marx**

While açai berry juice has the starring role in all of the incredible recipes you'll find in this book, a smoothie wouldn't be a smoothie without a good helping of fresh fruit. This chapter introduces you to the different fruits used in this book and guides you in selecting, storing, and preparing them to create the best açai smoothies possible.

First of all, you need to know that choosing "smoothie-ready" fruit can be difficult, especially if you base your choice on appearance alone. For example, a peach may appear to be ripe if it has a deep, rich color. However, there are several other, less-obvious attributes that are equally – or even more – important when selecting fruit, such as whether it has a fresh aroma, how heavy or dense it is, its resilience to the touch, and so on. It may seem a bit complicated at first, but in no time you'll become a fruit connoisseur and an expert at determining whether fruit is ripe and "smoothie-ready." What's more, I'm confident that the more you familiarize yourself with the fabulous array of fruit available in stores and markets today, the more you'll enjoy making refreshing, satisfying, and nourishing smoothies. Whether you stick to the recipes in this book or feel inspired to mix and match your own fruity ingredients, you'll be set to embark on a flavorful and healthy new adventure. I hope the following information helps you select the freshest ingredients from your favorite farmer's market or produce department, so you can sample the best that nature's bounty has to offer.

APRICOT

The apricot is a very sweet and juicy fruit with golden-orange skin and flesh and a single smooth stone. It is round or oblong and measures about two inches in diameter. Native to northern China, the apricot has been a food source since as early as 2200 BC. It is a rich source of Vitamin A, Vitamin C, potassium, iron, and fiber.

Selection

Look for apricots that are well colored and plump. Choose apricots that are fairly firm but yield slightly if you press them gently. Apricots that feel soft and juicy are fairly ripe and should be eaten right away. Apricots that are hard can be placed in a brown paper bag to ripen at room temperature for one or two days. Ripe apricots can be refrigerated in the crisper bin for up to a week. Wash apricots in cool water just before using them.

BANANA

According to Hindu legend, the banana was the forbidden fruit in the Garden of Eden – not the apple. It is also believed that bananas were widely cultivated in Asia and Oceania before recorded history. They were reportedly introduced to the New World by Spanish colonists in 1516. Bananas are an excellent source of Vitamin A, Vitamin B6, Vitamin C, and fiber, and they are regarded as one of nature's best energy sources. A banana is an ideal snack post-exercise, because it replaces potassium and other important nutrients that are often lost during strenuous activity.

Selection

Bananas are picked when they are green and continue to ripen (turning yellow) on the way to market. The riper a banana, the more yellow its skin, and the sweeter it is. When choosing bananas, look for ones that are yellow (and thus fully ripened). Don't worry if there's a bit of green – bananas that are yellow with green tips and/or green necks are ready to eat. Bananas that are still mostly green can be ripened at room temperature for two or three days. To accelerate the process, you can place them in a brown paper bag to ripen. Add an apple to the bag to make them ripen even more quickly. Once ripe, bananas can be stored at room temperature or refrigerated for a couple of days.

BLACKBERRY

The blackberry is a small, oblong berry that can be black, blue, or dark-red in color. Blackberries grow on brambles (thorny bushes), and the berries can be up to one inch long. They reach their peak flavor and availability from June through September, but you can sometimes find blackberries in supermarkets from November until April. The U.S. is the world's greatest blackberry producer. The berries are a great source of Vitamin C, fiber, and folate.

Selection

Look for blackberries that are plump and solid, with a bright, fresh appearance and rich color. Place the berries in a shallow container to prevent the ones on the bottom from getting crushed. Cover the container and store the blackberries in the crisper bin of the refrigerator for up to two days. Wash blackberries in cool water just before using them.

BLUEBERRY

The blueberry is native to North America. It is the second most popular berry in the United States, which produces 95% of the world's commercial crop of blueberries. Although it has existed for thousands of years, the blueberry was not cultivated until the turn of the 20th century. As the blueberry season progresses, they are available first in the southern states and then gradually move north. Blueberries are at their peak flavor from mid-April to late September, and they are an excellent source of Vitamin A, Vitamin C, and fiber.

Selection

Look for plump and firm blueberries with a dark-blue color and a silvery "bloom." (The "bloom" on blueberries is the powder that protects them from the sun. It cannot be rinsed off.) Do not choose blueberries that appear to be dull – it could be a sign that they are old.

Like blackberries, blueberries should be placed in a shallow pan and covered to avoid getting crushed; however, blueberries can be stored in the crisper bin for a longer time, from three to five days.

CANTALOUPE

Many people are surprised to learn that melons are in fact members of the cucumber family and that they grow on vines. The cantaloupe belongs to the muskmelon category of melons (which includes summer melons, such as the cantaloupe, and winter melons, such as the casaba or honeydew). Cantaloupes are an excellent source of Vitamin C.

Selection

Look for a cantaloupe that is unblemished and firm, with no soft spots. Find one that feels heaviest for its size. Smell the stem end of the cantaloupe to check for a fresh, melon aroma – if it doesn't have that melon aroma, the cantaloupe is not ripe. However, you can ripen a melon by placing it in a loosely closed brown paper bag. Wash cantaloupe in cool water and refrigerate it until you use it.

CHERRY

Cherries are small, round fruit that grow on trees in temperate areas of Europe, Asia, and the Americas. They can be red to black in color. There are several varieties of cherry, all of which fall into one of three categories: sweet, sour, or a combination of the two. It is believed that cherries originated in northeastern Asia and were spread throughout other temperate zones in prehistory by birds that ate cherries and later dropped the stones. Cherries are available from late May through early August and are a good source of Vitamin C and fiber.

Selection

Choose cherries with an attached stem and fruit that is dark, red, plump, and firm. Cherries can be stored in the crisper bin of the refrigerator for up to two days. Wash them in cool water just before using them.

KIWIFRUIT

The kiwifruit (or kiwi) grows on a vine and is about the size of a plum. The brown, fuzzy skin of kiwis envelops a luscious sweet-and-sour emerald-green pulp, which surrounds a cluster of edible black seeds. Kiwis are traced back to the Yangtze River valley in China, where they were first found in the 1600s. Known as the "Yantao" in China, its seeds were sent to New Zealand in 1906, and the fruit was called the "Chinese gooseberry." The Chinese gooseberry was then shipped to the U.S. in 1962, when it received its current name – "kiwifruit" – in honor of New Zealand's famous national bird. Kiwifruit is available all year and is rich in Vitamin C, Vitamin E, fiber, and potassium.

Selection

Look for kiwis that are light brown and that are firm but give slightly when pressed. They also should have a sweet aroma. Kiwis can be ripened at room temperature in three to five days or placed in a brown paper bag with an apple or banana to speed up the ripening process. Store ripe kiwis in the crisper bin for as long as three weeks.

MANGO

The mango has a slightly elongated shape. Its thin skin is waxy and smooth and can be green, red, orange, yellow or a combination of any of those colors. The aromatic, juicy pulp surrounds a hard inner pit. Mangoes were first cultivated in India and the Malay Archipelago as many as 4,000 years ago, and they were introduced to other tropical areas by European explorers in the 1700s and 1800s. They were first grown in the U.S. in the early 1900s. Mangoes are a rich source of beta carotene, vitamin C, potassium, and fiber.

Selection

Look for mangoes that are very fragrant and plump around the stem area and that give slightly when pressed. While the main color of the mango is not important, the best-flavored ones will have a yellow tinge when ripe. Mangoes can be ripened at room temperature; for accelerated ripening, they can be placed overnight in a brown paper bag along with an apple. Ripe mangoes can be stored in the crisper bin of the refrigerator for up to five days. Wash mangoes in cool water and dry well just before using.

PEACH

Peaches have been grown since prehistoric times. They were first cultivated in China and later introduced into Europe and Persia. It is believed that peaches were brought to the Americas by the Spaniards. There are several varieties of peaches available, and the various types are broken down into rough classifications. One type is the freestone, which has a pit that separates easily from the flesh of the peach. Freestone peaches are the type most often found in supermarkets, since they are easy to eat. Another variety of peach is the clingstone, which has a pit that is firmly attached to the fruit. Clingstone peaches are frequently canned. Peaches are ripe in summertime, reaching their peak flavor in August. They are a good source of Vitamin C.

Selection

Look for peaches that are relatively firm with a sweet aroma. As for color, pink blush on a peach indicates its variety, not its ripeness. Look for peaches with a fuzzy, creamy yellow skin, and avoid peaches with wrinkled skin or blemishes. The peach should not be soft, but it should yield gently when touched. Peaches can be ripened at room temperature – out of direct sun – until the skin yields slightly to the touch. Ripe peaches can be stored in the crisper bin of the refrigerator for up to five days. Store peaches in a single layer to avoid blemishing, and wash them in cool water just before using.

PINEAPPLE

The pineapple is a juicy tropical fruit with a sweet and tart flavor. It is native to Central and South America. Christopher Columbus discovered pineapples on the island of Guadeloupe in 1493 and brought them back to Spain. By the 1700s, pineapples were being cultivated in greenhouses throughout Europe. Pineapples are available year-round and are an excellent source of vitamin C.

Selection

Look for pineapples that are heavy and symmetrical in size. They also should have a brightly colored shell, fresh pineapple aroma, and a crown of crisp, fresh-looking green leaves. Avoid pineapples that are discolored or have any soft spots. To store a pineapple, remove the fruit from the shell and place it in an airtight container in the refrigerator for up to a week.

POMEGRANATE

The pomegranate is about the size of an orange, with a rind that varies in color from yellow-orange to deep crimson red or purple. The inside of the fruit contains seed sacs with pips that resemble corn kernels with reddish, translucent skin. These kernels hold the sweet juice of the pomegranate. Pomegranates are best in early fall but can often be found through early winter. Pomegranate is an excellent source of antioxidants and Vitamin C.

RASPBERRY

It is believed that wild red raspberries spread throughout Europe and Asia in prehistoric times. Because the wild berries were so delicious, raspberries were not cultivated in Europe until the 1600s. Raspberries cultivated in North America have origins from two groups: the red raspberry native to Europe and the wild red variety native to North America. Raspberries are rich in Vitamin C, fiber, and folate.

Selection

It is always best to buy raspberries in season, which usually starts in late June and lasts for four to six weeks. If you're fortunate to live near a local berry farm, take advantage and visit at the beginning of the season to get the best berries. Select raspberries that are large and plump, free of mold, and that have a bright, shiny, and uniform color. Avoid any raspberries that are mushy. If you purchase pre-packaged raspberries, carefully go through the batch and discard any berries that show signs of spoilage. After carefully washing raspberries, they should be stored in a shallow container to prevent the ones on bottom from being crushed. Cover the container and store in the crisper bin of the refrigerator for up to two days. Raspberries should be washed in a gentle stream of cool water just before using them.

STRAWBERRY

Strawberries have been traced back as far as 2,200 years ago, and they are known to have grown wild in the third century in Italy. In 1588, the first European settlers discovered strawberries in Virginia, and by the middle of the 19th century strawberries were being cultivated in many parts of North America. The fruit grows in groups of three on the stem of a very low plant. The color of a strawberry changes from greenish white to a lush flame red as the fruit ripens. Strawberries do not have a skin; they are covered by hundreds of tiny seeds. They are a very good source of Vitamin C and fiber.

Selection

Strawberries are at their peak in June and July. As with raspberries, those people fortunate enough to live near a strawberry farm should take advantage and pick their own berries to get the best of the crop. When picking strawberries, look for ones that are both plump and firm, and have a deep red color and bright green cap. Ripe fruit should have a sweet strawberry aroma. Strawberries should be washed and placed in a single layer in the crisper bin of the refrigerator. They can be stored in the refrigerator for up to two days. Wash strawberries (with their caps) in a gentle stream of cool water just before using them.

Freezing Fruit

To get the optimal consistency for a smoothie, use fresh fruit that has been frozen for at least 30 minutes. This also helps maintain smoothies at an ideal icy temperature. In addition, freezing fruit is an excellent way to store it for later use – which is especially useful when you know that a certain seasonal fruit will no longer be available. For example, you can purchase an ample quantity of blackberries or other seasonal fruits and freeze them so you can make your favorite smoothies any time of the year. Freezing is also a good option when you have fruit that is ripe but not needed immediately; freezing prevents over-ripening and lets you use the fruit at a later time.

Whether you need to freeze fruit for immediate use or longer-term storage, the basic preparation techniques are the same:

• To prepare cherries, apricots, and berries, place them in a colander, rinse with a gentle stream of cool water, and pat dry with a paper towel. Cherries and apricots should first be cut in half and their stones should be removed.

• To prepare peaches, first remove the stones and then cut the fruit into small pieces.

• To prepare a banana or kiwifruit, remove their skin. You can either slice them before freezing or freeze them whole and then slice them before use.

• To prepare mangoes and cantaloupe, remove their peels and their seeds or pits before cubing.

• To freeze a pineapple, cut off its crown and leaves, remove the outer layering and the core, and then cut the fruit into cubes.

Once properly prepared for freezing, place the fruit on a baking sheet lined with either freezer paper (plastic-coated side facing up) or non-stick aluminum foil. If neither is available, you can use waxed paper or parchment paper instead. Freeze the fruit for thirty minutes or longer to make it "smoothie-ready." If you are freezing fruit to use at a later date, follow the first step (thirty minutes or longer on the baking sheet) and then transfer the frozen pieces to a large, airtight plastic bag. The bag must be large enough to hold the pieces of fruit in a single layer. Most fruit can be kept in the freezer for up to two weeks without any loss of flavor, so be sure to label and mark the date on the bag.

How Much Fruit Should I Buy?

Use the table below to determine the quantity of fruit you need for your açai smoothies. The table gives you an estimate of how much fruit you will get once you have removed the skin, seeds, pit, core, and other unwanted bits. You can use the average weight per fruit or – to be more precise – weigh the fruit at the market before purchase.

Fruit	How Much To Buy	Average Weight	Number of Cups
Apricots	3	8 ounces	1 cup
Banana	1 large	10 ounces	1 cup
Blackberries	½ pint	6 ounces	1¼ cups
Blueberries	½ pint	8 ounces	1 cup
Cantaloupe	1 medium	3 pounds	5 cups
Cherries	19 to 20	8 ounces	1 cup
Kiwifruit	3	8 ounces	1 cup
Mango	1 medium	10 ounces	1 cup
Peach	1 medium	8 ounces	1 cup
Pineapple	1 medium	3 pounds	5½ cups
Raspberries	1 box	6 ounces	1¼ cups
Strawberries	7 to 8 medium	6 ounces	1 cup

Lessons for Life—How to Make an Açai Smoothie

The world is but a canvas to the imagination.
—**Henry David Thoreau**

The açai berry is rapidly becoming one of the most popular foods for health-conscious individuals and those looking for the fountain of youth. Now that you know a little more about the açai "superberry" and its potent punch of vitamins, minerals, and antioxidants, you may be ready to rush down to the market and load up on ingredients to make your first mouthwatering açai smoothies. Remember that açai juice must be refrigerated, so look for it where other specialty healthful refrigerated juices are displayed. If açai juice is not available, you can purchase açai fruit powder or açai smoothie packs online. These can be used as a substitute for fresh açai juice. Keep in mind, however, that using açai powder will eliminate liquid from the recipes. Follow the açai powder manufacturer's instructions for adding liquid to your smoothies. But before you get started, you need to make sure you have the right tools to make your açai berry creations at home.

The most important piece of equipment you'll need is a blender. You may already own one; if you don't, they are widely available at most department stores, hardware stores, kitchen shops, and on-line. (Alternatively, if you don't own a blender but do have a food processor, feel free to whip up your smoothies in that appliance – it works great.) When choosing a blender, the basic qualities you should look for are durability and ease of cleaning (for example, a removable bottom). Another advantageous feature is a secondary lid that can be removed easily to allow you to add ingredients while the blender is in use –

without creating a huge mess to clean up afterwards. Once you have the blender, the only remaining things needed are measuring cups and spoons, a spatula, and a healthy appetite!

Now that your private health club is equipped, you're well on your way to creating any of the 120 flavorful recipes in this book and discovering how incredible they taste and – better yet – how incredible they make you feel. As you try out the different recipes, remember these three essential (and easy) rules: 1) make sure to use the ripest fruits, 2) add the liquids first, followed by the fruits, and 3) add more fruit for a thicker smoothie or more açai berry juice for a thinner drink.

While I have prepared, tasted, and perfected each of the açai smoothie recipes in this book, feel free to get creative! Don't hesitate to substitute one or more ingredients to reflect your personal tastes. For example, in many of my recipes you'll find frozen juice cubes made with pomegranate juice. If you're more of a citrus fan, try using orange juice instead. The same principle applies if you are following my recipe for Açai, Blackberry, Cherry, and Banana Smoothie. If you can't find blackberries or just prefer a different taste, try preparing this unbelievably addictive and delectable smoothie with blueberries instead — it's perfectly fine! Combine açai berry juice with any variety of fruit and you are sure to create a drink that tastes heavenly… and healthy!

So, you've got the equipment on the counter, the açai berry juice in the refrigerator, and your favorite mix of fruit in the freezer… The fountain of youth is beckoning you – let the magic begin!

Açai Elixirs of Vitality — A Gift from the Amazon

⊱≼•≽⊰

❝ *Judge a tree from its fruit; not from the leaves.* **❞**
—Euripides

Ready to ride the latest health wave? It's time to prepare your first açai smoothie! You'll see just how easy it is to create impressively delicious and refreshing smoothies laden with unimaginable health benefits. After a few tries, you'll be ready to impress your friends by whipping up these healthful treats in just minutes.

In this chapter, you'll discover more than 40 scrumptious recipes that turn a simple serving of açai juice into a refreshing glassful of nutritious ecstasy. Mixed with fruits and other flavorful ingredients, açai is a pure, healthy indulgence. Get ready to be electrified when you try the intense flavors of an Açai, Peach, Raspberry, Apricot, and Raspberry Sorbet smoothie or an Açai, Blueberry, Strawberry, and Pomegranate smoothie.

Go ahead and try a few. Work your way through the whole chapter and then start experimenting with some of your own combinations – you're only limited by your imagination! Even Häagen-Dazs has joined the Açai revolution with its recently introduced new line of reserved sorbets, including Brazilian Açai Berry Sorbet. If it fits into your budget, consider adding a dollop of this new Acai source to one of the recipes found in this book in place of some or all the sorbets called for. With so many recipes at your disposal, you'll see how easy it is to create delicious, energy-boosting açai smoothies.

Açai, Apricot, Banana, and Peach

1 SERVING

1 cup (or more) açai juice
1 tablespoon (or to taste) honey
1 cup almost frozen diced apricots
½ cup almost frozen diced banana
½ cup almost frozen diced peach

Combine all the ingredients in a blender container in the order listed.

Cover the container; and then turn on the blender. Press the pulse button on its lowest blade-speed setting; and process until the ingredients are mostly blended.

Continue to mix on the highest blade-speed setting button until the mixture is smooth (it may be necessary to turn off the blender periodically to stir the mixture with a spoon, always working from the bottom up).

Turn off the blender. Scrape the smoothie into a glass.

Açai, Apricot, Banana, Mango, and Zesty Lemon Sorbet

꧁ꕤ꧂

1 SERVING

1 cup (or more) açai juice

1 tablespoon (or to taste) honey

1 cup almost frozen diced apricots

½ cup almost frozen diced banana

½ cup almost frozen diced mango

½ cup zesty lemon (or favorite flavor) sorbet, preferably Häagen-Dazs

Combine all the ingredients in a blender container in the order listed.

Cover the container; and then turn on the blender. Press the pulse button on its lowest blade-speed setting; and process until the ingredients are mostly blended.

Continue to mix on the highest blade-speed setting button until the mixture is smooth (it may be necessary to turn off the blender periodically to stir the mixture with a spoon, always working from the bottom up).

Turn off the blender. Scrape the smoothie into a glass.

Açai, Apricot, Cherry, Pineapple, and Pomegranate

1 SERVING

1 cup (or more) açai juice
1 tablespoon (or to taste) honey
1 cup almost frozen diced apricots
½ cup almost frozen cherries
½ cup almost frozen diced pineapple
½ cup frozen pomegranate juice cubes (see note)

Combine all the ingredients in a blender container in the order listed.

Cover the container; and then turn on the blender. Press the pulse button on its lowest blade-speed setting; and process until the ingredients are mostly blended.

Continue to mix on the highest blade-speed setting button until the mixture is smooth (it may be necessary to turn off the blender periodically to stir the mixture with a spoon, always working from the bottom up.)

Turn off the blender. Scrape the smoothie into a glass.

Note: To make frozen pomegranate juice cubes, pour pomegranate juice into plastic ice cube trays (preferably the easy-to-release variety). Freeze until completely firm; then unmold the cubes onto a cold pan or plate. Transfer the cubes to a re-sealable freezer bag and store them in the freezer. Use as needed.

Açai, Apricot, Peach, Banana, and Tropical Sorbet

1 SERVING

1 cup (or more) açai juice

1 tablespoon (or to taste) honey

1 cup almost frozen diced apricots

½ cup almost frozen diced peach

½ cup almost frozen diced banana

½ cup tropical (or favorite flavor) sorbet, preferably Häagen-Dazs

Combine all the ingredients in a blender container in the order listed.

Cover the container; and then turn on the blender. Press the pulse button on its lowest blade-speed setting; and process until the ingredients are mostly blended.

Continue to mix on the highest blade-speed setting button until the mixture is smooth (it may be necessary to turn off the blender periodically to stir the mixture with a spoon, always working from the bottom up).

Turn off the blender. Scrape the smoothie into a glass.

Açai, Apricot, Pineapple, Peach, and Peach Sorbet

1 SERVING

1 cup (or more) açai juice

1 tablespoon (or to taste) honey

1 cup almost frozen diced apricots

½ cup almost frozen diced pineapple

½ cup almost frozen diced peach

½ cup peach (or favorite flavor) sorbet, preferably Häagen-Dazs

Combine all the ingredients in a blender container in the order listed.

Cover the container; and then turn on the blender. Press the pulse button on its lowest blade-speed setting; and process until the ingredients are mostly blended.

Continue to mix on the highest blade-speed setting button until the mixture is smooth (it may be necessary to turn off the blender periodically to stir the mixture with a spoon, always working from the bottom up).

Turn off the blender. Scrape the smoothie into a glass.

Açai, Blueberry, and Raspberry

1 SERVING

1 cup (or more) açai juice
1 tablespoon (or to taste) honey
1 cup almost frozen blueberries
1 cup almost frozen raspberries

Combine all the ingredients in a blender container in the order listed.

Cover the container; and then turn on the blender. Press the pulse button on its lowest blade-speed setting; and process until the ingredients are mostly blended.

Continue to mix on the highest blade-speed setting button until the mixture is smooth (it may be necessary to turn off the blender periodically to stir the mixture with a spoon, always working from the bottom up).

Turn off the blender. Scrape the smoothie into a glass.

Açai, Blueberry, Blackberry, and Banana

❧

1 SERVING

1 cup (or more) açai juice
1 tablespoon (or to taste) honey
1 cup almost frozen blueberries
½ cup almost frozen blackberries
½ cup almost frozen diced banana

Combine all the ingredients in a blender container in the order listed.

Cover the container; and then turn on the blender. Press the pulse button on its lowest blade-speed setting; and process until the ingredients are mostly blended.

Continue to mix on the highest blade-speed setting button until the mixture is smooth (it may be necessary to turn off the blender periodically to stir the mixture with a spoon, always working from the bottom up).

Turn off the blender. Scrape the smoothie into a glass.

Açai, Blueberry, Cherry, Banana, and Coconut Sorbet

1 SERVING

1 cup (or more) açai juice

1 tablespoon (or to taste) honey

1 cup almost frozen blueberries

½ cup almost frozen cherries

½ cup almost frozen diced banana

½ cup coconut (or favorite flavor) sorbet, preferably Häagen-Dazs

Combine all the ingredients in a blender container in the order listed.

Cover the container; and then turn on the blender. Press the pulse button on its lowest blade-speed setting; and process until the ingredients are mostly blended.

Continue to mix on the highest blade-speed setting button until the mixture is smooth (it may be necessary to turn off the blender periodically to stir the mixture with a spoon, always working from the bottom up).

Turn off the blender. Scrape the smoothie into a glass.

Açai, Blueberry, Mango, and Apricot

❧

1 SERVING

1 cup (or more) açai juice
1 tablespoon (or to taste) honey
1 cup almost frozen blueberries
1 cup almost frozen diced mango
½ cup frozen apricot juice cubes (see note)

Combine all the ingredients in a blender container in the order listed.

Cover the container; and then turn on the blender. Press the pulse button on its lowest blade-speed setting; and process until the ingredients are mostly blended.

Continue to mix on the highest blade-speed setting button until the mixture is smooth (it may be necessary to turn off the blender periodically to stir the mixture with a spoon, always working from the bottom up).

Turn off the blender. Scrape the smoothie into a glass.

Note: To make frozen apricot juice cubes, pour apricot juice into plastic ice cube trays (preferably the easy-to-release variety). Freeze until completely firm; then unmold the cubes onto a cold pan or plate. Transfer the cubes to a re-sealable freezer bag and store them in the freezer. Use as needed.

Açai, Blueberry, Mango, Banana, and Mango Sorbet

1 SERVING

1 cup (or more) açai juice

1 tablespoon (or to taste) honey

1 cup almost frozen blueberries

½ cup almost frozen diced mango

½ cup almost frozen diced banana

½ cup mango (or favorite flavor) sorbet, preferably Häagen-Dazs

Combine all the ingredients in a blender container in the order listed.

Cover the container; and then turn on the blender. Press the pulse button on its lowest blade-speed setting; and process until the ingredients are mostly blended.

Continue to mix on the highest blade-speed setting button until the mixture is smooth (it may be necessary to turn off the blender periodically to stir the mixture with a spoon, always working from the bottom up).

Turn off the blender. Scrape the smoothie into a glass.

Açai, Blueberry, Pineapple, Kiwi, and Coconut Sorbet

1 SERVING

1 cup (or more) açai juice

1 tablespoon (or to taste) honey

1 cup almost frozen blueberries

½ cup almost frozen diced pineapple

½ cup almost frozen diced kiwi

½ cup coconut (or favorite flavor) sorbet,
preferably Häagen-Dazs

Combine all the ingredients in a blender container in the order listed.

Cover the container; and then turn on the blender. Press the pulse button on its lowest blade-speed setting; and process until the ingredients are mostly blended.

Continue to mix on the highest blade-speed setting button until the mixture is smooth (it may be necessary to turn off the blender periodically to stir the mixture with a spoon, always working from the bottom up).

Turn off the blender. Scrape the smoothie into a glass.

Açai, Blueberry, Pineapple, Peach, and Peach Sorbet

1 SERVING

1 cup (or more) açai juice

1 tablespoon (or to taste) honey

1 cup almost frozen blueberries

½ cup almost frozen diced pineapple

½ cup almost frozen diced peach

½ cup peach (or favorite flavor) sorbet,
  preferably Häagen-Dazs

Combine all the ingredients in a blender container in the order listed.

Cover the container; and then turn on the blender. Press the pulse button on its lowest blade-speed setting; and process until the ingredients are mostly blended.

Continue to mix on the highest blade-speed setting button until the mixture is smooth (it may be necessary to turn off the blender periodically to stir the mixture with a spoon, always working from the bottom up).

Turn off the blender. Scrape the smoothie into a glass.

Açai, Blueberry, Raspberry, and Apricot

❧

1 SERVING

1 cup (or more) açai juice
1 tablespoon (or to taste) honey
1 cup almost frozen blueberries
1 cup almost frozen raspberries
½ cup frozen apricot juice cubes (see note)

Combine all the ingredients in a blender container in the order listed.

Cover the container; and then turn on the blender. Press the pulse button on its lowest blade-speed setting; and process until the ingredients are mostly blended.

Continue to mix on the highest blade-speed setting button until the mixture is smooth (it may be necessary to turn off the blender periodically to stir the mixture with a spoon, always working from the bottom up).

Turn off the blender. Scrape the smoothie into a glass.

Note: To make frozen apricot juice cubes, pour apricot juice into plastic ice cube trays (preferably the easy-to-release variety). Freeze until completely firm; then unmold the cubes onto a cold pan or plate. Transfer the cubes to a re-sealable freezer bag and store them in the freezer. Use as needed.

Açai, Blueberry, Raspberry, Kiwi, and Pomegranate

1 SERVING

1 cup (or more) açai juice
1 tablespoon (or to taste) honey
1 cup almost frozen blueberries
½ cup almost frozen raspberries
½ cup almost frozen diced kiwi
½ cup frozen pomegranate juice cubes (see note)

Combine all the ingredients in a blender container in the order listed.

Cover the container; and then turn on the blender. Press the pulse button on its lowest blade-speed setting; and process until the ingredients are mostly blended.

Continue to mix on the highest blade-speed setting button until the mixture is smooth (it may be necessary to turn off the blender periodically to stir the mixture with a spoon, always working from the bottom up).

Turn off the blender. Scrape the smoothie into a glass.

Note: To make frozen pomegranate juice cubes, pour pomegranate juice into plastic ice cube trays (preferably the easy-to-release variety). Freeze until completely firm; then unmold the cubes onto a cold pan or plate. Transfer the cubes to a re-sealable freezer bag and store them in the freezer. Use as needed.

Açai, Blueberry, Strawberry, and Pomegranate

❦

1 SERVING

1 cup (or more) açai juice
1 tablespoon (or to taste) honey
1 cup almost frozen blueberries
1 cup almost frozen diced strawberries
½ cup frozen pomegranate juice cubes (see note)

Combine all the ingredients in a blender container in the order listed.

Cover the container; and then turn on the blender. Press the pulse button on its lowest blade-speed setting; and process until the ingredients are mostly blended.

Continue to mix on the highest blade-speed setting button until the mixture is smooth (it may be necessary to turn off the blender periodically to stir the mixture with a spoon, always working from the bottom up).

Turn off the blender. Scrape the smoothie into a glass.

Note: To make frozen pomegranate juice cubes, pour pomegranate juice into plastic ice cube trays (preferably the easy-to-release variety). Freeze until completely firm; then unmold the cubes onto a cold pan or plate. Transfer the cubes to a re-sealable freezer bag and store them in the freezer. Use as needed.

Açai, Blueberry, Strawberry, Banana, and Strawberry Sorbet

1 SERVING

1 cup (or more) açai juice

1 tablespoon (or to taste) honey

1 cup almost frozen blueberries

½ cup almost frozen diced strawberries

½ cup almost frozen diced banana

½ cup strawberry (or favorite flavor) sorbet, preferably Häagen-Dazs

Combine all the ingredients in a blender container in the order listed.

Cover the container; and then turn on the blender. Press the pulse button on its lowest blade-speed setting; and process until the ingredients are mostly blended.

Continue to mix on the highest blade-speed setting button until the mixture is smooth (it may be necessary to turn off the blender periodically to stir the mixture with a spoon, always working from the bottom up).

Turn off the blender. Scrape the smoothie into a glass.

Açai, Blueberry, Strawberry, Pomegranate, and Coconut Sorbet

1 SERVING

1 cup (or more) açai juice

1 tablespoon (or to taste) honey

1 cup almost frozen blueberries

1 cup almost frozen diced strawberries

½ cup frozen pomegranate juice cubes (see note)

½ cup coconut (or favorite flavor) sorbet, preferably Häagen-Dazs

Combine all the ingredients in a blender container in the order listed.

Cover the container; and then turn on the blender. Press the pulse button on its lowest blade-speed setting; and process until the ingredients are mostly blended.

Continue to mix on the highest blade-speed setting button until the mixture is smooth (it may be necessary to turn off the blender periodically to stir the mixture with a spoon, always working from the bottom up).

Turn off the blender. Scrape the smoothie into a glass.

Note: To make frozen pomegranate juice cubes, pour pomegranate juice into plastic ice cube trays (preferably the easy-to-release variety). Freeze until completely firm; then unmold the cubes onto a cold pan or plate. Transfer the cubes to a re-sealable freezer bag and store them in the freezer. Use as needed.

Açai, Cherry, Peach, Banana, and Peach Sorbet

1 SERVING

1 cup (or more) açai juice

1 tablespoon (or to taste) honey

1 cup almost frozen cherries

½ cup almost frozen diced peach

½ cup almost frozen diced banana

½ cup peach (or favorite flavor) sorbet,
  preferably Häagen-Dazs

Combine all the ingredients in a blender container in the order listed.

Cover the container; and then turn on the blender. Press the pulse button on its lowest blade-speed setting; and process until the ingredients are mostly blended.

Continue to mix on the highest blade-speed setting button until the mixture is smooth (it may be necessary to turn off the blender periodically to stir the mixture with a spoon, always working from the bottom up).

Turn off the blender. Scrape the smoothie into a glass.

Açai, Cherry, Strawberry, and Pomegranate

1 SERVING

1 cup (or more) açai juice
1 tablespoon (or to taste) honey
1 cup almost frozen cherries
1 cup almost frozen diced strawberries
½ cup frozen pomegranate juice cubes (see note)

Combine all the ingredients in a blender container in the order listed.

Cover the container; and then turn on the blender. Press the pulse button on its lowest blade-speed setting; and process until the ingredients are mostly blended.

Continue to mix on the highest blade-speed setting button until the mixture is smooth (it may be necessary to turn off the blender periodically to stir the mixture with a spoon, always working from the bottom up).

Turn off the blender. Scrape the smoothie into a glass.

Note: To make frozen pomegranate juice cubes, pour pomegranate juice into plastic ice cube trays (preferably the easy-to-release variety). Freeze until completely firm; then unmold the cubes onto a cold pan or plate. Transfer the cubes to a re-sealable freezer bag and store them in the freezer. Use as needed.

Açai, Mango, Blackberry, Peach, Pomegranate, and Peach Sorbet

1 SERVING

1 cup (or more) açai juice

1 tablespoon (or to taste) honey

1 cup almost frozen diced mango

½ cup almost frozen blackberries

½ cup almost frozen diced peach

½ cup frozen pomegranate juice cubes (see note)

½ cup peach (or favorite flavor) sorbet,
  preferably Häagen-Dazs

Combine all the ingredients in a blender container in the order listed.

Cover the container; and then turn on the blender. Press the pulse button on its lowest blade-speed setting; and process until the ingredients are mostly blended.

Continue to mix on the highest blade-speed setting button until the mixture is smooth (it may be necessary to turn off the blender periodically to stir the mixture with a spoon, always working from the bottom up).

Turn off the blender. Scrape the smoothie into a glass.

Note: To make frozen pomegranate juice cubes, pour pomegranate juice into plastic ice cube trays (preferably the easy-to-release variety). Freeze until completely firm; then unmold the cubes onto a cold pan or plate. Transfer the cubes to a re-sealable freezer bag and store them in the freezer. Use as needed.

Açai, Mango, Blackberry, Pineapple, Pomegranate, and Mango Sorbet

1 SERVING

1 cup (or more) açai juice

1 tablespoon (or to taste) honey

1 cup almost frozen diced mango

½ cup almost frozen blackberries

½ cup almost frozen diced pineapple

½ cup frozen pomegranate juice cubes (see note)

½ cup mango (or favorite flavor) sorbet, preferably Häagen-Dazs

Combine all the ingredients in a blender container in the order listed.

Cover the container; and then turn on the blender. Press the pulse button on its lowest blade-speed setting; and process until the ingredients are mostly blended.

Continue to mix on the highest blade-speed setting button until the mixture is smooth (it may be necessary to turn off the blender periodically to stir the mixture with a spoon, always working from the bottom up).

Turn off the blender. Scrape the smoothie into a glass.

Note: To make frozen pomegranate juice cubes, pour pomegranate juice into plastic ice cube trays (preferably the easy-to-release variety). Freeze until completely firm; then unmold the cubes onto a cold pan or plate. Transfer the cubes to a re-sealable freezer bag and store them in the freezer. Use as needed.

Açai, Mango, Blueberry, Peach, Pomegranate, and Zesty Lemon Sorbet

1 SERVING

1 cup (or more) açai juice
1 tablespoon (or to taste) honey
1 cup almost frozen diced mango
½ cup almost frozen blueberries
½ cup almost frozen diced peach
½ cup frozen pomegranate juice cubes (see note)
½ cup zesty lemon (or favorite flavor) sorbet, preferably Häagen-Dazs

Combine all the ingredients in a blender container in the order listed.

Cover the container; and then turn on the blender. Press the pulse button on its lowest blade-speed setting; and process until the ingredients are mostly blended.

Continue to mix on the highest blade-speed setting button until the mixture is smooth (it may be necessary to turn off the blender periodically to stir the mixture with a spoon, always working from the bottom up).

Turn off the blender. Scrape the smoothie into a glass.

Note: To make frozen pomegranate juice cubes, pour pomegranate juice into plastic ice cube trays (preferably the easy-to-release variety). Freeze until completely firm; then unmold the cubes onto a cold pan or plate. Transfer the cubes to a re-sealable freezer bag and store them in the freezer. Use as needed.

Açai, Mango, Blueberry, Pineapple, Pomegranate, and Raspberry Sorbet

1 SERVING

1 cup (or more) açai juice

1 tablespoon (or to taste) honey

1 cup almost frozen diced mango

½ cup almost frozen blueberries

½ cup almost frozen diced pineapple

½ cup frozen pomegranate juice cubes (see note)

½ cup raspberry (or favorite flavor) sorbet,
  preferably Häagen-Dazs

Combine all the ingredients in a blender container in the order listed.

Cover the container; and then turn on the blender. Press the pulse button on its lowest blade-speed setting; and process until the ingredients are mostly blended.

Continue to mix on the highest blade-speed setting button until the mixture is smooth (it may be necessary to turn off the blender periodically to stir the mixture with a spoon, always working from the bottom up).

Turn off the blender. Scrape the smoothie into a glass.

Note: To make frozen pomegranate juice cubes, pour pomegranate juice into plastic ice cube trays (preferably the easy-to-release variety). Freeze until completely firm; then unmold the cubes onto a cold pan or plate. Transfer the cubes to a re-sealable freezer bag and store them in the freezer. Use as needed.

Açai, Mango, Raspberry, Pineapple, Pomegranate, and Tropical Sorbet

1 SERVING

1 cup (or more) açai juice

1 tablespoon (or to taste) honey

1 cup almost frozen diced mango

½ cup almost frozen raspberries

½ cup almost frozen diced pineapple

½ cup frozen pomegranate juice cubes (see note)

½ cup tropical (or favorite flavor) sorbet, preferably Häagen-Dazs

Combine all the ingredients in a blender container in the order listed.

Cover the container; and then turn on the blender. Press the pulse button on its lowest blade-speed setting; and process until the ingredients are mostly blended.

Continue to mix on the highest blade-speed setting button until the mixture is smooth (it may be necessary to turn off the blender periodically to stir the mixture with a spoon, always working from the bottom up).

Turn off the blender. Scrape the smoothie into a glass.

Note: To make frozen pomegranate juice cubes, pour pomegranate juice into plastic ice cube trays (preferably the easy-to-release variety). Freeze until completely firm; then unmold the cubes onto a cold pan or plate. Transfer the cubes to a re-sealable freezer bag and store them in the freezer. Use as needed.

Açai, Mango, Strawberry, Pineapple, Pomegranate, and Strawberry Sorbet

1 SERVING

1 cup (or more) açai juice

1 tablespoon (or to taste) honey

1 cup almost frozen diced mango

½ cup almost frozen diced strawberries

½ cup almost frozen diced pineapple

½ cup frozen pomegranate juice cubes (see note)

½ cup strawberry (or favorite flavor) sorbet,
   preferably Häagen-Dazs

Combine all the ingredients in a blender container in the order listed.

Cover the container; and then turn on the blender. Press the pulse button on its lowest blade-speed setting; and process until the ingredients are mostly blended.

Continue to mix on the highest blade-speed setting button until the mixture is smooth (it may be necessary to turn off the blender periodically to stir the mixture with a spoon, always working from the bottom up).

Turn off the blender. Scrape the smoothie into a glass.

Note: To make frozen pomegranate juice cubes, pour pomegranate juice into plastic ice cube trays (preferably the easy-to-release variety). Freeze until completely firm; then unmold the cubes onto a cold pan or plate. Transfer the cubes to a re-sealable freezer bag and store them in the freezer. Use as needed.

Açai, Peach, Banana, and Raspberry

1 SERVING

1 cup (or more) açai juice
1 tablespoon (or to taste) honey
1 cup almost frozen diced peach
½ cup almost frozen diced banana
½ cup almost frozen raspberries

Combine all the ingredients in a blender container in the order listed.

Cover the container; and then turn on the blender. Press the pulse button on its lowest blade-speed setting; and process until the ingredients are mostly blended.

Continue to mix on the highest blade-speed setting button until the mixture is smooth (it may be necessary to turn off the blender periodically to stir the mixture with a spoon, always working from the bottom up).

Turn off the blender. Scrape the smoothie into a glass.

Açai, Peach, Mango, Apricot, and Tropical Sorbet

1 SERVING

1 cup (or more) açai juice

1 tablespoon (or to taste) honey

1 cup almost frozen diced peach

1 cup almost frozen diced mango

½ cup frozen apricot juice cubes (see note)

½ cup tropical (or favorite flavor) sorbet, preferably Häagen-Dazs

Combine all the ingredients in a blender container in the order listed.

Cover the container; and then turn on the blender. Press the pulse button on its lowest blade-speed setting; and process until the ingredients are mostly blended.

Continue to mix on the highest blade-speed setting button until the mixture is smooth (it may be necessary to turn off the blender periodically to stir the mixture with a spoon, always working from the bottom up).

Turn off the blender. Scrape the smoothie into a glass.

Note: To make frozen apricot ice cubes, pour apricot juice into plastic ice cube trays (preferably the easy-to-release variety). Freeze until completely firm; then unmold the cubes onto a cold pan or plate. Transfer the cubes to a re-sealable freezer bag and store them in the freezer. Use as needed.

Açai, Peach, Pineapple, Apricot, and Peach Sorbet

1 SERVING

1 cup (or more) açai juice
1 tablespoon (or to taste) honey
1 cup almost frozen diced peach
1 cup almost frozen diced pineapple
½ cup frozen apricot juice cubes (see note)
½ cup peach (or favorite flavor) sorbet, preferably Häagen-Dazs

Combine all the ingredients in a blender container in the order listed.

Cover the container; and then turn on the blender. Press the pulse button on its lowest blade-speed setting; and process until the ingredients are mostly blended.

Continue to mix on the highest blade-speed setting button until the mixture is smooth (it may be necessary to turn off the blender periodically to stir the mixture with a spoon, always working from the bottom up).

Turn off the blender. Scrape the smoothie into a glass.

Note: To make frozen apricot ice cubes, pour apricot juice into plastic ice cube trays (preferably the easy-to-release variety). Freeze until completely firm; then unmold the cubes onto a cold pan or plate. Transfer the cubes to a re-sealable freezer bag and store them in the freezer. Use as needed.

Açai, Peach, Pineapple, Banana, and Coconut Sorbet

1 SERVING

1 cup (or more) açai juice
1 tablespoon (or to taste) honey
1 cup almost frozen diced peach
½ cup almost frozen diced pineapple
½ cup almost frozen diced banana
½ cup coconut (or favorite flavor) sorbet, preferably Häagen-Dazs

Combine all the ingredients in a blender container in the order listed.

Cover the container; and then turn on the blender. Press the pulse button on its lowest blade-speed setting; and process until the ingredients are mostly blended.

Continue to mix on the highest blade-speed setting button until the mixture is smooth (it may be necessary to turn off the blender periodically to stir the mixture with a spoon, always working from the bottom up).

Turn off the blender. Scrape the smoothie into a glass.

Açai, Peach, Raspberry, Apricot, and Raspberry Sorbet

1 SERVING

1 cup (or more) açai uice

1 tablespoon (or to taste) honey

1 cup almost frozen diced peach

1 cup almost frozen raspberries

½ cup frozen apricot juice cubes (see note)

½ cup raspberry (or favorite flavor) sorbet, preferably Häagen-Dazs

Combine all the ingredients in a blender container in the order listed.

Cover the container; and then turn on the blender. Press the pulse button on its lowest blade-speed setting; and process until the ingredients are mostly blended.

Continue to mix on the highest blade-speed setting button until the mixture is smooth (it may be necessary to turn off the blender periodically to stir the mixture with a spoon, always working from the bottom up).

Turn off the blender. Scrape the smoothie into a glass.

Note: To make frozen apricot ice cubes, pour apricot juice into plastic ice cube trays (preferably the easy-to-release variety). Freeze until completely firm; then unmold the cubes onto a cold pan or plate. Transfer the cubes to a re-sealable freezer bag and store them in the freezer. Use as needed.

Açai, Pineapple, Blueberry, and Banana

❧⚜☙

1 SERVING

1 cup (or more) açai juice
1 tablespoon (or to taste) honey
1 cup almost frozen diced pineapple
½ cup almost frozen blueberries
½ cup almost frozen diced banana

Combine all the ingredients in a blender container in the order listed.

Cover the container; and then turn on the blender. Press the pulse button on its lowest blade-speed setting; and process until the ingredients are mostly blended.

Continue to mix on the highest blade-speed setting button until the mixture is smooth (it may be necessary to turn off the blender periodically to stir the mixture with a spoon, always working from the bottom up).

Turn off the blender. Scrape the smoothie into a glass.

Açai, Pineapple, Cherry, and Banana

1 SERVING

1 cup (or more) açai juice
1 tablespoon (or to taste) honey
1 cup almost frozen diced pineapple
½ cup almost frozen cherries
½ cup almost frozen diced banana

Combine all the ingredients in a blender container in the order listed.

Cover the container; and then turn on the blender. Press the pulse button on its lowest blade-speed setting; and process until the ingredients are mostly blended.

Continue to mix on the highest blade-speed setting button until the mixture is smooth (it may be necessary to turn off the blender periodically to stir the mixture with a spoon, always working from the bottom up).

Turn off the blender. Scrape the smoothie into a glass.

Açai, Raspberry, and Banana

1 SERVING

1 cup (or more) açai juice
1 tablespoon (or to taste) honey
1 cup almost frozen raspberries
1 cup almost frozen diced banana

Combine all the ingredients in a blender container in the order listed.

Cover the container; and then turn on the blender. Press the pulse button on its lowest blade-speed setting; and process until the ingredients are mostly blended.

Continue to mix on the highest blade-speed setting button until the mixture is smooth (it may be necessary to turn off the blender periodically to stir the mixture with a spoon, always working from the bottom up)

Turn off the blender. Scrape the smoothie into a glass.

Açai, Strawberry, Banana, and Raspberry

1 SERVING

1 cup (or more) açai juice
1 tablespoon (or to taste) honey
1 cup almost frozen diced strawberries
½ cup almost frozen diced banana
½ cup almost frozen raspberries

Combine all the ingredients in a blender container in the order listed.

Cover the container; and then turn on the blender. Press the pulse button on its lowest blade-speed setting; and process until the ingredients are mostly blended.

Continue to mix on the highest blade-speed setting button until the mixture is smooth (it may be necessary to turn off the blender periodically to stir the mixture with a spoon, always working from the bottom up).

Turn off the blender. Scrape the smoothie into a glass.

Açai, Strawberry, Blackberry, and Banana

1 SERVING

1 cup (or more) açai juice
1 tablespoon (or to taste) honey
1 cup almost frozen diced strawberries
½ cup almost frozen blackberries
½ cup almost frozen diced banana

Combine all the ingredients in a blender container in the order listed.

Cover the container; and then turn on the blender. Press the pulse button on its lowest blade-speed setting; and process until the ingredients are mostly blended.

Continue to mix on the highest blade-speed setting button until the mixture is smooth (it may be necessary to turn off the blender periodically to stir the mixture with a spoon, always working from the bottom up).

Turn off the blender. Scrape the smoothie into a glass.

Açai, Strawberry, Blueberry, Pomegranate, and Strawberry Sorbet

1 SERVING

1 cup (or more) açai juice

1 tablespoon (or to taste) honey

1 cup almost frozen diced strawberries

1 cup almost frozen blueberries

½ cup frozen pomegranate juice cubes (see note)

½ cup strawberry (or favorite flavor) sorbet, preferably Häagen-Dazs

Combine all the ingredients in a blender container in the order listed.

Cover the container; and then turn on the blender. Press the pulse button on its lowest blade-speed setting; and process until the ingredients are mostly blended.

Continue to mix on the highest blade-speed setting button until the mixture is smooth (it may be necessary to turn off the blender periodically to stir the mixture with a spoon, always working from the bottom up).

Turn off the blender. Scrape the smoothie into a glass.

Note: To make frozen pomegranate ice cubes, pour pomegranate juice into plastic ice cube trays (preferably the easy-to-release variety). Freeze until completely firm; then unmold the cubes onto a cold pan or plate. Transfer the cubes to a re-sealable freezer bag and store them in the freezer. Use as needed.

Açai, Strawberry, Mango, and Pomegranate

1 SERVING

1 cup (or more) açai juice
1 tablespoon (or to taste) honey
1 cup almost frozen diced strawberries
1 cup almost frozen diced mango
½ cup frozen pomegranate juice cubes (see note)

Combine all the ingredients in a blender container in the order listed.

Cover the container; and then turn on the blender. Press the pulse button on its lowest blade-speed setting; and process until the ingredients are mostly blended.

Continue to mix on the highest blade-speed setting button until the mixture is smooth (it may be necessary to turn off the blender periodically to stir the mixture with a spoon, always working from the bottom up).

Turn off the blender. Scrape the smoothie into a glass.

Note: To make frozen pomegranate ice cubes, pour pomegranate juice into plastic ice cube trays (preferably the easy-to-release variety). Freeze until completely firm; then unmold the cubes onto a cold pan or plate. Transfer the cubes to a re-sealable freezer bag and store them in the freezer. Use as needed.

Açai, Strawberry, Mango, Orange, and Mango Sorbet

1 SERVING

1 cup (or more) açai juice

1 tablespoon (or to taste) honey

1 cup almost frozen diced strawberries

1 cup almost frozen diced mango

½ cup frozen orange juice cubes (see note)

½ cup mango (or favorite flavor) sorbet, preferably Häagen-Dazs

Combine all the ingredients in a blender container in the order listed.

Cover the container; and then turn on the blender. Press the pulse button on its lowest blade-speed setting; and process until the ingredients are mostly blended.

Continue to mix on the highest blade-speed setting button until the mixture is smooth (it may be necessary to turn off the blender periodically to stir the mixture with a spoon, always working from the bottom up).

Turn off the blender. Scrape the smoothie into a glass.

Note: To make frozen orange ice cubes, pour orange juice into plastic ice cube trays (preferably the easy-to-release variety). Freeze until completely firm; then unmold the cubes onto a cold pan or plate. Transfer the cubes to a re-sealable freezer bag and store them in the freezer. Use as needed.

Açai and Yogurt
— Fountains of Youth

❝*Yoghurt is very good for the stomach, the lumbar regions, appendicitis and apotheosis.* ❞
—**Eugene Ionesco,
Romanian writer (1912-1994)**

While any açai smoothie is a delicious and nutritious delight, the creaminess of yogurt makes smoothies especially indulgent. If you're looking for that little extra something to add to your magical rainforest creations, look no further than this collection of mouth-watering yogurt recipes. These açai yogurt smoothies are rich and creamy antioxidant-fortified sensations. You may even be able to find açai-flavored yogurt in your local health foods store or supermarket -- you can use it to get an extra kick of açai sweetness in your smoothies.

Choose from mouthwatering combinations like *Açai, Blackberry, Blueberry, Cherry, and Vanilla Yogurt; Acai, Pineapple, Banana, Apricot, and Pineapple Yogurt; and Açai, Blackberry, Pineapple, Pomegranate, and Vanilla Yogurt.* Sipping one of these luscious beverages is a delectable way to get the antioxidants, vitamins and nutrients you need to stay young, healthy, and vibrant!

Açai, Apricot, Banana, Peach, and Peach Yogurt

1 SERVING

1 cup (or more) açai juice

1 tablespoon (or to taste) honey

1 cup almost frozen diced apricots

½ cup almost frozen diced banana

½ cup almost frozen diced peach

1 (6-ounce) container peach (or favorite flavor) whole milk, low fat, or nonfat yogurt

Combine all the ingredients in a blender container in the order listed.

Cover the container; and then turn on the blender. Press the pulse button on its lowest blade-speed setting; and process until the ingredients are mostly blended.

Continue to mix on the highest blade-speed setting button until the mixture is smooth (it may be necessary to turn off the blender periodically to stir the mixture with a spoon, always working from the bottom up).

Turn off the blender. Scrape the smoothie into a glass.

Açai, Apricot, Cantaloupe, Pineapple, and Vanilla Yogurt

1 SERVING

1 cup (or more) açai juice
1 tablespoon (or to taste) honey
1 cup almost frozen diced apricots
½ cup almost frozen diced cantaloupe
½ cup almost frozen diced pineapple
1 (6-ounce) container vanilla (or favorite flavor) whole milk, low fat, or nonfat yogurt

Combine all the ingredients in a blender container in the order listed.

Cover the container; and then turn on the blender. Press the pulse button on its lowest blade-speed setting; and process until the ingredients are mostly blended.

Continue to mix on the highest blade-speed setting button until the mixture is smooth (it may be necessary to turn off the blender periodically to stir the mixture with a spoon, always working from the bottom up).

Turn off the blender. Scrape the smoothie into a glass.

Açai, Apricot, Mango, Pomegranate, and Mango Yogurt

1 SERVING

1 cup (or more) açai juice

1 tablespoon (or to taste) honey

1 cup almost frozen diced apricots

1 cup almost frozen diced mango

½ cup frozen pomegranate juice cubes (see note)

1 (6-ounce) container mango (or favorite flavor) whole milk, low fat, or nonfat yogurt

Combine all the ingredients in a blender container in the order listed.

Cover the container; and then turn on the blender. Press the pulse button on its lowest blade-speed setting; and process until the ingredients are mostly blended.

Continue to mix on the highest blade-speed setting button until the mixture is smooth (it may be necessary to turn off the blender periodically to stir the mixture with a spoon, always working from the bottom up).

Turn off the blender. Scrape the smoothie into a glass.

Note: To make frozen pomegranate juice cubes, pour pomegranate juice into plastic ice cube trays (preferably the easy-to-release variety). Freeze until completely firm; then unmold the cubes onto a cold pan or plate. Transfer the cubes to a re-sealable freezer bag and store them in the freezer. Use as needed.

Açai, Apricot, Pineapple, Banana, and Apricot Yogurt

1 SERVING

1 cup (or more) açai juice
1 tablespoon (or to taste) honey
1 cup almost frozen diced apricots
½ cup almost frozen diced pineapple
½ cup almost frozen diced banana
1 (6-ounce) container apricot (or favorite flavor) whole milk, low fat, or nonfat yogurt

Combine all the ingredients in a blender container in the order listed.

Cover the container; and then turn on the blender. Press the pulse button on its lowest blade-speed setting; and process until the ingredients are mostly blended.

Continue to mix on the highest blade-speed setting button until the mixture is smooth (it may be necessary to turn off the blender periodically to stir the mixture with a spoon, always working from the bottom up).

Turn off the blender. Scrape the smoothie into a glass.

Açai, Banana, Pineapple, and Vanilla Yogurt

1 SERVING

1 cup (or more) açai juice

1 tablespoon (or to taste) honey

1 cup almost frozen diced banana

1 cup almost frozen diced pineapple

1 (6-ounce) container vanilla (or favorite flavor) whole milk, low fat, or nonfat yogurt

Combine all the ingredients in a blender container in the order listed.

Cover the container; and then turn on the blender. Press the pulse button on its lowest blade-speed setting; and process until the ingredients are mostly blended.

Continue to mix on the highest blade-speed setting button until the mixture is smooth (it may be necessary to turn off the blender periodically to stir the mixture with a spoon, always working from the bottom up).

Turn off the blender. Scrape the smoothie into a glass.

Açai, Blackberry, Blueberry, Cherry, and Vanilla Yogurt

1 SERVING

1 cup (or more) açai juice
1 tablespoon (or to taste) honey
1 cup almost frozen blackberries
½ cup almost frozen blueberries
½ cup almost frozen cherries
1 (6-ounce) container vanilla (or favorite flavor)
  whole milk, low fat, or nonfat yogurt

Combine all the ingredients in a blender container in the order listed.

Cover the container; and then turn on the blender. Press the pulse button on its lowest blade-speed setting; and process until the ingredients are mostly blended.

Continue to mix on the highest blade-speed setting button until the mixture is smooth (it may be necessary to turn off the blender periodically to stir the mixture with a spoon, always working from the bottom up).

Turn off the blender. Scrape the smoothie into a glass.

Açai, Blackberry, Mango, Banana, and Mango Yogurt

1 SERVING

1 cup (or more) açai juice

1 tablespoon (or to taste) honey

1 cup almost frozen blackberries

½ cup almost frozen diced mango

½ cup almost frozen diced banana

1 (6-ounce) container mango (or favorite flavor) whole milk, low fat, or nonfat yogurt

Combine all the ingredients in a blender container in the order listed.

Cover the container; and then turn on the blender. Press the pulse button on its lowest blade-speed setting; and process until the ingredients are mostly blended.

Continue to mix on the highest blade-speed setting button until the mixture is smooth (it may be necessary to turn off the blender periodically to stir the mixture with a spoon, always working from the bottom up).

Turn off the blender. Scrape the smoothie into a glass.

Açai, Blackberry, Pineapple, Pomegranate, and Vanilla Yogurt

1 SERVING

1 cup (or more) açai juice
1 tablespoon (or to taste) honey
¾ cup almost frozen blackberries
¾ cup almost frozen diced pineapple
½ cup frozen pomegranate juice cubes (see note)
1 (6-ounce) container vanilla (or favorite flavor) whole milk, low fat, or nonfat yogurt

Combine all the ingredients in a blender container in the order listed.

Cover the container; and then turn on the blender. Press the pulse button on its lowest blade-speed setting; and process until the ingredients are mostly blended.

Continue to mix on the highest blade-speed setting button until the mixture is smooth (it may be necessary to turn off the blender periodically to stir the mixture with a spoon, always working from the bottom up).

Turn off the blender. Scrape the smoothie into a glass.

Note: To make frozen pomegranate juice cubes, pour pomegranate juice into plastic ice cube trays (preferably the easy-to-release variety). Freeze until completely firm; then unmold the cubes onto a cold pan or plate. Transfer the cubes to a re-sealable freezer bag and store them in the freezer. Use as needed.

Açai, Blueberry, Pineapple, Peach, Pomegranate, and Peach Yogurt

1 SERVING

1 cup (or more) açai juice

1 tablespoon (or to taste) honey

¾ cup almost frozen blueberries

½ cup almost frozen diced pineapple

½ cup almost frozen diced peach

½ cup frozen pomegranate juice cubes (see note)

1 (6-ounce) container peach (or favorite flavor) whole milk, low fat, or nonfat yogurt

Combine all the ingredients in a blender container in the order listed.

Cover the container; and then turn on the blender. Press the pulse button on its lowest blade-speed setting; and process until the ingredients are mostly blended.

Continue to mix on the highest blade-speed setting button until the mixture is smooth (it may be necessary to turn off the blender periodically to stir the mixture with a spoon, always working from the bottom up).

Turn off the blender. Scrape the smoothie into a glass.

Note: To make frozen pomegranate juice cubes, pour pomegranate juice into plastic ice cube trays (preferably the easy-to-release variety). Freeze until completely firm; then unmold the cubes onto a cold pan or plate. Transfer the cubes to a re-sealable freezer bag and store them in the freezer. Use as needed.

Açai, Cantaloupe, Banana, Orange, and Vanilla Yogurt

1 SERVING

1 cup (or more) açai juice
1 tablespoon (or to taste) honey
1 cup almost frozen diced cantaloupe
1 cup almost frozen diced banana
½ cup frozen orange juice cubes
1 (6-ounce) container vanilla (or favorite flavor) whole milk, low fat, or nonfat yogurt

Combine all the ingredients in a blender container in the order listed.

Cover the container; and then turn on the blender. Press the pulse button on its lowest blade-speed setting; and process until the ingredients are mostly blended.

Continue to mix on the highest blade-speed setting button until the mixture is smooth (it may be necessary to turn off the blender periodically to stir the mixture with a spoon, always working from the bottom up).

Turn off the blender. Scrape the smoothie into a glass.

Note: To make frozen orange juice cubes, pour orange juice into plastic ice cube trays (preferably the easy-to-release variety). Freeze until completely firm; then unmold the cubes onto a cold pan or plate. Transfer the cubes to a re-sealable freezer bag and store them in the freezer. Use as needed.

Açai, Cantaloupe, Blueberry, Banana, and Blueberry Yogurt

1 SERVING

1 cup (or more) açai juice
1 tablespoon (or to taste) honey
1 cup almost frozen diced cantaloupe
½ cup almost frozen blueberries
½ cup almost frozen diced banana
1 (6-ounce) container blueberry (or favorite flavor) whole milk, low fat, or nonfat yogurt

Combine all the ingredients in a blender container in the order listed.

Cover the container; and then turn on the blender. Press the pulse button on its lowest blade-speed setting; and process until the ingredients are mostly blended.

Continue to mix on the highest blade-speed setting button until the mixture is smooth (it may be necessary to turn off the blender periodically to stir the mixture with a spoon, always working from the bottom up).

Turn off the blender. Scrape the smoothie into a glass.

Açai, Cantaloupe, Peach, Banana, and Peach Yogurt

1 SERVING

1 cup (or more) açai juice
1 tablespoon (or to taste) honey
1 cup almost frozen diced cantaloupe
½ cup almost frozen diced peach
½ cup almost frozen diced banana
1 (6-ounce) container peach (or favorite flavor) whole milk, low fat, or nonfat yogurt

Combine all the ingredients in a blender container in the order listed.

Cover the container; and then turn on the blender. Press the pulse button on its lowest blade-speed setting; and process until the ingredients are mostly blended.

Continue to mix on the highest blade-speed setting button until the mixture is smooth (it may be necessary to turn off the blender periodically to stir the mixture with a spoon, always working from the bottom up).

Turn off the blender. Scrape the smoothie into a glass.

Açai, Cherry, Raspberry, Banana, and Raspberry Açai Berry Yogurt

1 SERVING

1 cup (or more) açai juice
1 tablespoon (or to taste) honey
1 cup almost frozen cherries
½ cup almost frozen raspberries
½ cup almost frozen diced banana
1 (6-ounce) container raspberry açai berry (or favorite flavor) whole milk, low fat, or nonfat yogurt

Combine all the ingredients in a blender container in the order listed.

Cover the container; and then turn on the blender. Press the pulse button on its lowest blade-speed setting; and process until the ingredients are mostly blended.

Continue to mix on the highest blade-speed setting button until the mixture is smooth (it may be necessary to turn off the blender periodically to stir the mixture with a spoon, always working from the bottom up).

Turn off the blender. Scrape the smoothie into a glass.

Açai, Cherry, Pineapple, Banana, and Cherry Yogurt

1 SERVING

1 cup (or more) açai juice
1 tablespoon (or to taste) honey
1 cup almost frozen cherries
½ cup almost frozen diced pineapple
½ cup almost frozen diced banana
1 (6-ounce) container cherry (or favorite flavor) whole milk, low fat, or nonfat yogurt

Combine all the ingredients in a blender container in the order listed.

Cover the container; and then turn on the blender. Press the pulse button on its lowest blade-speed setting; and process until the ingredients are mostly blended.

Continue to mix on the highest blade-speed setting button until the mixture is smooth (it may be necessary to turn off the blender periodically to stir the mixture with a spoon, always working from the bottom up).

Turn off the blender. Scrape the smoothie into a glass.

Açai, Pineapple, Blueberry, Banana, and Raspberry Açai Berry Yogurt

1 SERVING

1 cup (or more) açai juice

1 tablespoon (or to taste) honey

1 cup almost frozen diced pineapple

½ cup almost frozen blueberries

½ cup almost frozen diced banana

1 (6-ounce) container raspberry açai berry (or favorite flavor) whole milk, low fat, or nonfat yogurt

Combine all the ingredients in a blender container in the order listed.

Cover the container; and then turn on the blender. Press the pulse button on its lowest blade-speed setting; and process until the ingredients are mostly blended.

Continue to mix on the highest blade-speed setting button until the mixture is smooth (it may be necessary to turn off the blender periodically to stir the mixture with a spoon, always working from the bottom up).

Turn off the blender. Scrape the smoothie into a glass.

Açai, Pineapple, Cantaloupe, Banana, and Pineapple Yogurt

1 SERVING

1 cup (or more) açai juice
1 tablespoon (or to taste) honey
1 cup almost frozen diced pineapple
½ cup almost frozen diced cantaloupe
½ cup almost frozen diced banana
1 (6-ounce) container pineapple (or favorite flavor)
whole milk, low fat, or nonfat yogurt

Combine all the ingredients in a blender container in the order listed.

Cover the container; and then turn on the blender. Press the pulse button on its lowest blade-speed setting; and process until the ingredients are mostly blended.

Continue to mix on the highest blade-speed setting button until the mixture is smooth (it may be necessary to turn off the blender periodically to stir the mixture with a spoon, always working from the bottom up).

Turn off the blender. Scrape the smoothie into a glass.

Açai, Pineapple, Mango, Apricot, and Mango Yogurt

1 SERVING

1 cup (or more) açai juice

1 tablespoon (or to taste) honey

¾ cup almost frozen diced pineapple

¾ cup almost frozen diced mango

½ cup frozen apricot juice cubes (see note)

1 (6-ounce) container mango (or favorite flavor) whole milk, low fat, or nonfat yogurt

Combine all the ingredients in a blender container in the order listed.

Cover the container; and then turn on the blender. Press the pulse button on its lowest blade-speed setting; and process until the ingredients are mostly blended.

Continue to mix on the highest blade-speed setting button until the mixture is smooth (it may be necessary to turn off the blender periodically to stir the mixture with a spoon, always working from the bottom up).

Turn off the blender. Scrape the smoothie into a glass.

Note: To make frozen apricot ice cubes, pour apricot juice into plastic ice cube trays (preferably the easy-to-release variety). Freeze until completely firm; then unmold the cubes onto a cold pan or plate. Transfer the cubes to a re-sealable freezer bag and store them in the freezer. Use as needed.

Açai, Pineapple, Peach, Apricot, and Pineapple Yogurt

1 SERVING

1 cup (or more) açai juice
1 tablespoon (or to taste) honey
¾ cup almost frozen diced pineapple
¾ cup almost frozen diced peach
½ cup frozen apricot juice cubes (see note)
1 (6-ounce) container pineapple (or favorite flavor) whole milk, low fat, or nonfat yogurt

Combine all the ingredients in a blender container in the order listed.

Cover the container; and then turn on the blender. Press the pulse button on its lowest blade-speed setting; and process until the ingredients are mostly blended.

Continue to mix on the highest blade-speed setting button until the mixture is smooth (it may be necessary to turn off the blender periodically to stir the mixture with a spoon, always working from the bottom up).

Turn off the blender. Scrape the smoothie into a glass.

Note: To make frozen apricot ice cubes, pour apricot juice into plastic ice cube trays (preferably the easy-to-release variety). Freeze until completely firm; then unmold the cubes onto a cold pan or plate. Transfer the cubes to a re-sealable freezer bag and store them in the freezer. Use as needed.

Açai, Raspberry, Banana, and Blueberry Yogurt

1 SERVING

1 cup (or more) açai juice

1 tablespoon (or to taste) honey

1 cup almost frozen raspberries

1 cup almost frozen diced banana

1 (6-ounce) container blueberry (or favorite flavor) whole milk, low fat, or nonfat yogurt

Combine all the ingredients in a blender container in the order listed.

Cover the container; and then turn on the blender. Press the pulse button on its lowest blade-speed setting; and process until the ingredients are mostly blended.

Continue to mix on the highest blade-speed setting button until the mixture is smooth (it may be necessary to turn off the blender periodically to stir the mixture with a spoon, always working from the bottom up).

Turn off the blender. Scrape the smoothie into a glass.

Açai, Banana, Pineapple, Pomegranate, and Strawberry Yogurt

❧

1 SERVING

1 cup (or more) açai juice

1 tablespoon (or to taste) honey

¾ cup almost frozen diced banana

¾ cup almost frozen diced pineapple

½ cup frozen pomegranate juice cubes (see note)

1 (6-ounce) container strawberry (or favorite flavor) whole milk, low fat, or nonfat yogurt

Combine all the ingredients in a blender container in the order listed.

Cover the container; and then turn on the blender. Press the pulse button on its lowest blade-speed setting; and process until the ingredients are mostly blended.

Continue to mix on the highest blade-speed setting button until the mixture is smooth (it may be necessary to turn off the blender periodically to stir the mixture with a spoon, always working from the bottom up).

Turn off the blender. Scrape the smoothie into a glass.

Note: To make pomegranate ice cubes, pour pomegranate juice into plastic ice cube trays (preferably the easy-to-release variety). Freeze until completely firm; then unmold the cubes onto a cold pan or plate. Transfer the cubes to a re-sealable freezer bag and store them in the freezer. Use as needed.

Açai, Blueberry, Apricot, Mango, and Blueberry Yogurt

1 SERVING

1 cup (or more) açai juice
1 tablespoon (or to taste) honey
1 cup almost frozen blueberries
½ cup almost frozen diced apricots
½ cup almost frozen diced mango
1 (6-ounce) container blueberry (or favorite flavor) whole milk, low fat, or nonfat yogurt

Combine all the ingredients in a blender container in the order listed.

Cover the container; and then turn on the blender. Press the pulse button on its lowest blade-speed setting; and process until the ingredients are mostly blended.

Continue to mix on the highest blade-speed setting button until the mixture is smooth (it may be necessary to turn off the blender periodically to stir the mixture with a spoon, always working from the bottom up).

Turn off the blender. Scrape the smoothie into a glass.

Açai, Blueberry, Blackberry, Banana, and Blueberry Yogurt

1 SERVING

1 cup (or more) açai juice
1 tablespoon (or to taste) honey
1 cup almost frozen blueberries
½ cup almost frozen blackberries
½ cup almost frozen diced banana
1 (6-ounce) container blueberry (or favorite flavor)
   whole milk, low fat, or nonfat yogurt

Combine all the ingredients in a blender container in the order listed.

Cover the container; and then turn on the blender. Press the pulse button on its lowest blade-speed setting; and process until the ingredients are mostly blended.

Continue to mix on the highest blade-speed setting button until the mixture is smooth (it may be necessary to turn off the blender periodically to stir the mixture with a spoon, always working from the bottom up).

Turn off the blender. Scrape the smoothie into a glass.

Açai, Cherry, Banana, Apricot, and Cherry Yogurt

∽✦∽

1 SERVING

1 cup (or more) açai juice

1 tablespoon (or to taste) honey

1 cup almost frozen cherries

½ cup almost frozen diced banana

½ cup almost frozen diced apricots

1 (6-ounce) container cherry (or favorite flavor)
whole milk, low fat, or nonfat yogurt

Combine all the ingredients in a blender container in the order listed.

Cover the container; and then turn on the blender. Press the pulse button on its lowest blade-speed setting; and process until the ingredients are mostly blended.

Continue to mix on the highest blade-speed setting button until the mixture is smooth (it may be necessary to turn off the blender periodically to stir the mixture with a spoon, always working from the bottom up).

Turn off the blender. Scrape the smoothie into a glass.

Açai, Cherry, Strawberry, Banana, and Strawberry Yogurt

1 SERVING

1 cup (or more) açai juice

1 tablespoon (or to taste) honey

1 cup almost frozen cherries

½ cup almost frozen diced strawberries

½ cup almost frozen diced banana

1 (6-ounce) container strawberry (or favorite flavor) whole milk, low fat, or nonfat yogurt

Combine all the ingredients in a blender container in the order listed.

Cover the container; and then turn on the blender. Press the pulse button on its lowest blade-speed setting; and process until the ingredients are mostly blended.

Continue to mix on the highest blade-speed setting button until the mixture is smooth (it may be necessary to turn off the blender periodically to stir the mixture with a spoon, always working from the bottom up).

Turn off the blender. Scrape the smoothie into a glass.

Açai, Mango, Peach, Kiwi, and Raspberry Açai Berry Yogurt

❊

1 SERVING

1 cup (or more) açai juice

1 tablespoon (or to taste) honey

1 cup almost frozen diced mango

½ cup almost frozen diced peach

½ cup almost frozen diced kiwi

1 (6-ounce) container raspberry açai berry
(or favorite flavor) whole milk, low fat, or
nonfat yogurt

Combine all the ingredients in a blender container in
the order listed.

Cover the container; and then turn on the blender.
Press the pulse button on its lowest blade-speed setting;
and process until the ingredients are mostly blended.

Continue to mix on the highest blade-speed setting
button until the mixture is smooth (it may be
necessary to turn off the blender periodically to stir
the mixture with a spoon, always working from the
bottom up).

Turn off the blender. Scrape the smoothie into a glass.

Açai, Peach, Pineapple, Banana, and Pineapple Yogurt

❦

1 SERVING

1 cup (or more) açai juice

1 tablespoon (or to taste) honey

1 cup almost frozen diced peach

½ cup almost frozen diced pineapple

½ cup almost frozen diced banana

1 (6-ounce) container pineapple (or favorite flavor) whole milk, low fat, or nonfat yogurt

Combine all the ingredients in a blender container in the order listed.

Cover the container; and then turn on the blender. Press the pulse button on its lowest blade-speed setting; and process until the ingredients are mostly blended.

Continue to mix on the highest blade-speed setting button until the mixture is smooth (it may be necessary to turn off the blender periodically to stir the mixture with a spoon, always working from the bottom up).

Turn off the blender. Scrape the smoothie into a glass.

Açai, Peach, Raspberry, Banana, and Raspberry Açai Berry Yogurt

1 SERVING

1 cup (or more) açai juice
1 tablespoon (or to taste) honey
1 cup almost frozen diced peach
½ cup almost frozen diced banana
½ cup almost frozen raspberries
1 (6-ounce) container raspberry açai berry (or favorite flavor) whole milk, low fat, or nonfat yogurt

Combine all the ingredients in a blender container in the order listed.

Cover the container; and then turn on the blender. Press the pulse button on its lowest blade-speed setting; and process until the ingredients are mostly blended.

Continue to mix on the highest blade-speed setting button until the mixture is smooth (it may be necessary to turn off the blender periodically to stir the mixture with a spoon, always working from the bottom up).

Turn off the blender. Scrape the smoothie into a glass.

Açai, Pineapple, Banana, Apricot, and Pineapple Yogurt

1 SERVING

½ cup (or more) açai juice

1 tablespoon (or to taste) honey

1 cup almost frozen diced pineapple

½ cup almost frozen diced banana

½ cup almost frozen diced apricots

1 (6-ounce) container pineapple (or favorite flavor) whole milk, low fat, or nonfat yogurt

Combine all the ingredients in a blender container in the order listed.

Cover the container; and then turn on the blender. Press the pulse button on its lowest blade-speed setting; and process until the ingredients are mostly blended.

Continue to mix on the highest blade-speed setting button until the mixture is smooth (it may be necessary to turn off the blender periodically to stir the mixture with a spoon, always working from the bottom up).

Turn off the blender. Scrape the smoothie into a glass.

Açai, Pineapple, Mango, Pomegranate, and Pineapple Yogurt

❧

1 SERVING

½ cup (or more) açai juice

1 tablespoon (or to taste) honey

¾ cup almost frozen diced pineapple

¾ cup almost frozen diced mango

½ cup frozen pomegranate juice cubes (see note)

1 (6-ounce) container pineapple (or favorite flavor) whole milk, low fat, or nonfat yogurt

Combine all the ingredients in a blender container in the order listed.

Cover the container; and then turn on the blender. Press the pulse button on its lowest blade-speed setting; and process until the ingredients are mostly blended.

Continue to mix on the highest blade-speed setting button until the mixture is smooth (it may be necessary to turn off the blender periodically to stir the mixture with a spoon, always working from the bottom up).

Turn off the blender. Scrape the smoothie into a glass.

Note: To make pomegranate ice cubes, pour pomegranate juice into plastic ice cube trays (preferably the easy-to-release variety). Freeze until completely firm; then unmold the cubes onto a cold pan or plate. Transfer the cubes to a re-sealable freezer bag and store them in the freezer. Use as needed.

Açai, Pineapple, Raspberry, Banana, and Raspberry Açai Berry Yogurt

1 SERVING

1 cup (or more) açai juice

1 tablespoon (or to taste) honey

1 cup almost frozen diced pineapple

½ cup almost frozen raspberries

½ cup almost frozen diced banana

1 (6-ounce) container raspberry açai berry (or favorite flavor) whole milk, low fat, or nonfat yogurt

Combine all the ingredients in a blender container in the order listed.

Cover the container; and then turn on the blender. Press the pulse button on its lowest blade-speed setting; and process until the ingredients are mostly blended.

Continue to mix on the highest blade-speed setting button until the mixture is smooth (it may be necessary to turn off the blender periodically to stir the mixture with a spoon, always working from the bottom up).

Turn off the blender. Scrape the smoothie into a glass.

Açai, Raspberry, Banana, Cherry, and Cherry Yogurt

1 SERVING

1 cup (or more) açai juice

1 tablespoon (or to taste) honey

1 cup almost frozen raspberries

½ cup almost frozen diced banana

½ cup almost frozen cherries

1 (6-ounce) container cherry (or favorite flavor) whole milk, low fat, or nonfat yogurt

Combine all the ingredients in a blender container in the order listed.

Cover the container; and then turn on the blender. Press the pulse button on its lowest blade-speed setting; and process until the ingredients are mostly blended.

Continue to mix on the highest blade-speed setting button until the mixture is smooth (it may be necessary to turn off the blender periodically to stir the mixture with a spoon, always working from the bottom up).

Turn off the blender. Scrape the smoothie into a glass.

Açai, Raspberry, Blackberry, Banana, and Raspberry Açai Berry Yogurt

❧

1 SERVING

- 1 cup (or more) açai juice
- 1 tablespoon (or to taste) honey
- 1 cup almost frozen raspberries
- ½ cup almost frozen blackberries
- ½ cup almost frozen diced banana
- 1 (6-ounce) container raspberry açai berry (or favorite flavor) whole milk, low fat, or nonfat yogurt

Combine all the ingredients in a blender container in the order listed.

Cover the container; and then turn on the blender. Press the pulse button on its lowest blade-speed setting; and process until the ingredients are mostly blended.

Continue to mix on the highest blade-speed setting button until the mixture is smooth (it may be necessary to turn off the blender periodically to stir the mixture with a spoon, always working from the bottom up).

Turn off the blender. Scrape the smoothie into a glass.

Açai, Raspberry, Cantaloupe, Banana, and Vanilla Yogurt

٭

1 cup (or more) açai juice

1 tablespoon (or to taste) honey

1 cup almost frozen raspberries

½ cup almost frozen diced cantaloupe

½ cup almost frozen diced banana

1 (6-ounce) container vanilla (or favorite flavor) whole milk, low fat, or nonfat yogurt

Combine all the ingredients in a blender container in the order listed.

Cover the container; and then turn on the blender. Press the pulse button on its lowest blade-speed setting; and process until the ingredients are mostly blended.

Continue to mix on the highest blade-speed setting button until the mixture is smooth (it may be necessary to turn off the blender periodically to stir the mixture with a spoon, always working from the bottom up).

Turn off the blender. Scrape the smoothie into a glass.

Açai, Raspberry, Cherry, Pomegranate, and Cherry Yogurt

꩜

1 SERVING

1 cup (or more) açai juice

1 tablespoon (or to taste) honey

1 cup almost frozen raspberries

½ cup almost frozen cherries

½ cup frozen pomegranate juice cubes (see note)

1 (6-ounce) container cherry (or favorite flavor) whole milk, low fat, or nonfat yogurt

Combine all the ingredients in a blender container in the order listed.

Cover the container; and then turn on the blender. Press the pulse button on its lowest blade-speed setting; and process until the ingredients are mostly blended.

Continue to mix on the highest blade-speed setting button until the mixture is smooth (it may be necessary to turn off the blender periodically to stir the mixture with a spoon, always working from the bottom up).

Turn off the blender. Scrape the smoothie into a glass.

Note: To make pomegranate ice cubes, pour pomegranate juice into plastic ice cube trays (preferably the easy-to-release variety). Freeze until completely firm; then unmold the cubes onto a cold pan or plate. Transfer the cubes to a re-sealable freezer bag and store them in the freezer. Use as needed.

Açai, Strawberry, Apricot, Banana, and Apricot Yogurt

1 SERVING

1 cup (or more) açai juice

1 tablespoon (or to taste) honey

1 cup almost frozen diced strawberries

½ cup almost frozen diced apricots

½ cup almost frozen diced banana

1 (6-ounce) container apricot (or favorite flavor) whole milk, low fat, or nonfat yogurt

Combine all the ingredients in a blender container in the order listed.

Cover the container; and then turn on the blender. Press the pulse button on its lowest blade-speed setting; and process until the ingredients are mostly blended.

Continue to mix on the highest blade-speed setting button until the mixture is smooth (it may be necessary to turn off the blender periodically to stir the mixture with a spoon, always working from the bottom up).

Turn off the blender. Scrape the smoothie into a glass.

Açai, Strawberry, Apricot, Cantaloupe, and Strawberry Yogurt

1 SERVING

1 cup (or more) açai juice

1 tablespoon (or to taste) honey

¾ cup almost frozen diced strawberries

½ cup almost frozen diced apricots

½ cup almost frozen diced cantaloupe

½ cup frozen apricot juice cubes (see note)

1 (6-ounce) container strawberry (or favorite flavor) whole milk, low fat, or nonfat yogurt

Combine all the ingredients in a blender container in the order listed.

Cover the container; and then turn on the blender. Press the pulse button on its lowest blade-speed setting; and process until the ingredients are mostly blended.

Continue to mix on the highest blade-speed setting button until the mixture is smooth (it may be necessary to turn off the blender periodically to stir the mixture with a spoon, always working from the bottom up).

Turn off the blender. Scrape the smoothie into a glass.

Note: To make apricot ice cubes, pour apricot juice into plastic ice cube trays (preferably the easy-to-release variety). Freeze until completely firm; then unmold the cubes onto a cold pan or plate. Transfer the cubes to a re-sealable freezer bag and store them in the freezer. Use as needed.

Açai, Strawberry, Blueberry, Banana, and Vanilla Yogurt

1 SERVING

1 cup (or more) açai juice

1 tablespoon (or to taste) honey

1 cup almost frozen diced strawberries

½ cup almost frozen blueberries

½ cup almost frozen diced banana

1 (6-ounce) container vanilla (or favorite flavor)
whole milk, low fat, or nonfat yogurt

Combine all the ingredients in a blender container in the order listed.

Cover the container; and then turn on the blender. Press the pulse button on its lowest blade-speed setting; and process until the ingredients are mostly blended.

Continue to mix on the highest blade-speed setting button until the mixture is smooth (it may be necessary to turn off the blender periodically to stir the mixture with a spoon, always working from the bottom up).

Turn off the blender. Scrape the smoothie into a glass.

Açai, Strawberry, Cherry, Pomegranate, and Strawberry Yogurt

❦

1 SERVING

1 cup (or more) açai juice

1 tablespoon (or to taste) honey

¾ cup almost frozen diced strawberries

¾ cup almost frozen cherries

½ cup frozen pomegranate juice cubes (see note)

1 (6-ounce) container strawberry (or favorite flavor) whole milk, low fat, or nonfat yogurt

Combine all the ingredients in a blender container in the order listed.

Cover the container; and then turn on the blender. Press the pulse button on its lowest blade-speed setting; and process until the ingredients are mostly blended.

Continue to mix on the highest blade-speed setting button until the mixture is smooth (it may be necessary to turn off the blender periodically to stir the mixture with a spoon, always working from the bottom up).

Turn off the blender. Scrape the smoothie into a glass.

Note: To make pomegranate ice cubes, pour pomegranate juice into plastic ice cube trays (preferably the easy-to-release variety). Freeze until completely firm; then unmold the cubes onto a cold pan or plate. Transfer the cubes to a re-sealable freezer bag and store them in the freezer. Use as needed.

Açai, Strawberry, Pineapple, Banana, and Strawberry Yogurt

1 SERVING

1 cup (or more) açai juice

1 tablespoon (or to taste) honey

1 cup almost frozen diced strawberries

½ cup almost frozen diced pineapple

½ cup almost frozen diced banana

1 (6-ounce) container strawberry (or favorite flavor) whole milk, low fat, or nonfat yogurt

Combine all the ingredients in a blender container in the order listed.

Cover the container; and then turn on the blender. Press the pulse button on its lowest blade-speed setting; and process until the ingredients are mostly blended.

Continue to mix on the highest blade-speed setting button until the mixture is smooth (it may be necessary to turn off the blender periodically to stir the mixture with a spoon, always working from the bottom up).

Turn off the blender. Scrape the smoothie into a glass.

Açai, Strawberry, Raspberry, Pomegranate, and Vanilla Yogurt

❧❧❧

1 SERVING

1 cup (or more) açai juice

1 tablespoon (or to taste) honey

1 cup almost frozen diced strawberries

¾ cup almost frozen raspberries

½ cup frozen pomegranate juice cubes (see note)

1 (6-ounce) container vanilla (or favorite flavor) whole milk, low fat, or nonfat yogurt

Combine all the ingredients in a blender container in the order listed.

Cover the container; and then turn on the blender. Press the pulse button on its lowest blade-speed setting; and process until the ingredients are mostly blended.

Continue to mix on the highest blade-speed setting button until the mixture is smooth (it may be necessary to turn off the blender periodically to stir the mixture with a spoon, always working from the bottom up).

Turn off the blender. Scrape the smoothie into a glass.

Note: To make pomegranate ice cubes, pour pomegranate juice into plastic ice cube trays (preferably the easy-to-release variety). Freeze until completely firm; then unmold the cubes onto a cold pan or plate. Transfer the cubes to a re-sealable freezer bag and store them in the freezer. Use as needed.

Health Insurance: Açaí +

꧁꧂

❝ *We are indeed much more than what we eat, but what we eat can nevertheless help us to be much more than what we are.* ❞

—**Adelle Davis**

Açaí is a powerful and delicious package of vitamins, minerals, fiber, and so much more. But even the most perfect of foods can benefit from an all-star supporting cast. Health additives such as soy bean products, flaxseed oil, and wheat germ are great ways to enhance the textures, flavors, and nutritional content of your açaí smoothies.

You're sure to feel like a kid again with the energy boost you'll get from yummy and healthful smoothie recipes like *Açaí, Flaxseed, Raspberry, Pineapple, Blueberry,* and *Vanilla Soy Yogurt; Açaí, Mango, Cantaloupe, Pineapple, Vanilla Soymilk, Mango Soy Yogurt,* and *Granola;* and *Açaí, Mango, Cherry, Pineapple, Vanilla Soymilk, Mango Soy Yogurt,* and *Wheat Germ.*

You can also experiment with these recipes to discover which additives you like best and try adding them to some of the other great açaí smoothie recipes in this book. They're 100% natural – just like the açaí berry – so they're a great way to get that extra little charge of important nutrients that keep you going strong throughout the day.

Açai, Apricot, Mango, Pineapple, Pomegranate, Apricot Mango Soy Yogurt, and Granola

1 serving

1 cup (or more) açai juice

1 tablespoon (or to taste) honey

1 cup almost frozen diced apricots

½ cup almost frozen diced mango

½ cup almost frozen diced pineapple

½ cup frozen pomegranate juice cubes (see note)

1 (6-ounce) container apricot mango (or favorite flavor) soy yogurt

2 to 4 tablespoons favorite granola

Combine the açai juice, honey, apricots, mango, pineapple, pomegranate juice cubes, and apricot mango soy yogurt in a blender container in the order listed.

Cover the container; and then turn on the blender. Press the pulse button on its lowest blade-speed setting; and process until the ingredients are mostly blended.

Continue to mix on the highest blade-speed setting button until the mixture is smooth (it may be necessary to turn off the blender periodically to stir the mixture with a spoon, always working from the bottom up).

Turn off the blender. Add the granola and mix until just blended. Scrape the smoothie into a glass.

Note: To make pomegranate ice cubes, pour pomegranate juice into plastic ice cube trays (preferably the easy-to-release variety). Freeze until completely firm; then unmold the cubes onto a cold pan or plate. Transfer the cubes to a re-sealable freezer bag and store them in the freezer. Use as needed.

Açai, Apricot, Pineapple, Banana, Apricot Soy Yogurt, and Wheat Germ

1 SERVING

1 cup (or more) açai juice

1 tablespoon (or to taste) honey

1 cup almost frozen diced apricots

½ cup almost frozen diced pineapple

½ cup almost frozen diced banana

1 (6-ounce) container apricot (or favorite flavor) soy yogurt

2 to 4 tablespoons wheat germ (see note)

Combine the açai juice, honey, apricots, pineapple, banana, and apricot soy yogurt in a blender container in the order listed.

Cover the container; and then turn on the blender. Press the pulse button on its lowest blade-speed setting; and process until the ingredients are mostly blended.

Continue to mix on the highest blade-speed setting button until the mixture is smooth (it may be necessary to turn off the blender periodically to stir the mixture with a spoon, always working from the bottom up).

Turn off the blender. Add the wheat germ and mix until just blended. Scrape the smoothie into a glass.

Note: To toast wheat germ, spread ¼ cup (or more) wheat germ in a single layer on a rimmed cookie sheet. Bake in a preheated 350 degrees F oven for about 5 minutes.

Açai, Banana, Pineapple, and Vanilla Soymilk

꧁ꕥ꧂

1 SERVING

1 cup (or more) açai juice
1 tablespoon (or to taste) honey
1 cup almost frozen diced banana
1 cup almost frozen diced pineapple
½ cup frozen vanilla soymilk cubes (see note)

Combine all the ingredients in a blender container in the order listed.

Cover the container; and then turn on the blender. Press the pulse button on its lowest blade-speed setting; and process until the ingredients are mostly blended.

Continue to mix on the highest blade-speed setting button until the mixture is smooth (it may be necessary to turn off the blender periodically to stir the mixture with a spoon, always working from the bottom up).

Turn off the blender. Scrape the smoothie into a glass.

Note: To make soymilk ice cubes, pour soymilk into plastic ice cube trays (preferably the easy-to-release variety). Freeze until completely firm; then unmold the cubes onto a cold pan or plate. Transfer the cubes to a re-sealable freezer bag and store them in the freezer. Use as needed.

Açai, Blackberry, Banana, Apricot, Blackberry Soy Yogurt, and Wheat Germ

1 SERVING

½ cup (or more) açai juice

1 tablespoon (or to taste) honey

1 cup almost frozen blackberries

½ cup almost frozen diced banana

½ cup almost frozen diced apricots

1 (6-ounce) container blackberry (or favorite flavor) soy yogurt

2 to 4 tablespoons wheat germ (see note)

Combine the açai juice, honey, blackberries, banana, apricots, and blackberry soy yogurt in a blender container in the order listed.

Cover the container; and then turn on the blender. Press the pulse button on its lowest blade-speed setting; and process until the ingredients are mostly blended.

Continue to mix on the highest blade-speed setting button until the mixture is smooth (it may be necessary to turn off the blender periodically to stir the mixture with a spoon, always working from the bottom up).

Turn off the blender. Add the wheat germ and mix until just blended. Scrape the smoothie into a glass.

Note: To toast wheat germ, spread ? cup (or more) wheat germ in a single layer on a rimmed cookie sheet. Bake in a preheated 350 degrees F oven for about 5 minutes.

Açaí, Blackberry, Cherry, Strawberry, Vanilla Soymilk, and Strawberry Soy Yogurt

1 SERVING

1 cup (or more) açaí juice

1 tablespoon (or to taste) honey

1 cup almost frozen blackberries

½ cup almost frozen cherries

½ cup almost frozen diced strawberries

½ cup frozen vanilla soymilk cubes (see note)

1 (6-ounce) container strawberry (or favorite flavor) soy yogurt

Combine all the ingredients in a blender container in the order listed.

Cover the container; and then turn on the blender. Press the pulse button on its lowest blade-speed setting; and process until the ingredients are mostly blended.

Continue to mix on the highest blade-speed setting button until the mixture is smooth (it may be necessary to turn off the blender periodically to stir the mixture with a spoon, always working from the bottom up).

Turn off the blender. Scrape the smoothie into a glass.

Note: To make soymilk ice cubes, pour soymilk into plastic ice cube trays (preferably the easy-to-release variety). Freeze until completely firm; then unmold the cubes onto a cold pan or plate. Transfer the cubes to a re-sealable freezer bag and store them in the freezer. Use as needed.

Açai, Blackberry, Mango, Apricot, Vanilla Soymilk, and Blackberry Soy Yogurt

1 SERVING

1 cup (or more) açai juice

1 tablespoon (or to taste) honey

1 cup almost frozen blackberries

½ cup almost frozen diced mango

½ cup almost frozen diced apricots

2 tablespoons frozen vanilla soymilk cubes (see note)

1 (6-ounce) container blackberry (or favorite flavor) soy yogurt

Combine all the ingredients in a blender container in the order listed.

Cover the container; and then turn on the blender. Press the pulse button on its lowest blade-speed setting; and process until the ingredients are mostly blended.

Continue to mix on the highest blade-speed setting button until the mixture is smooth (it may be necessary to turn off the blender periodically to stir the mixture with a spoon, always working from the bottom up).

Turn off the blender. Scrape the smoothie into a glass.

Note: To make soymilk ice cubes, pour soymilk into plastic ice cube trays (preferably the easy-to-release variety). Freeze until completely firm; then unmold the cubes onto a cold pan or plate. Transfer the cubes to a re-sealable freezer bag and store them in the freezer. Use as needed.

Açai, Blackberry, Peach, Banana, Tofu, and Vanilla Soymilk

1 SERVING

1 cup (or more) açai juice
1 tablespoon (or to taste) honey
1 cup almost frozen blackberries
½ cup almost frozen diced peach
½ cup almost frozen diced banana
½ cup firm tofu, cut into chunks
¼ cup frozen vanilla soymilk cubes (see note)

Combine all the ingredients in a blender container in the order listed.

Cover the container; and then turn on the blender. Press the pulse button on its lowest blade-speed setting; and process until the ingredients are mostly blended.

Continue to mix on the highest blade-speed setting button until the mixture is smooth (it may be necessary to turn off the blender periodically to stir the mixture with a spoon, always working from the bottom up).

Turn off the blender. Scrape the smoothie into a glass.

Note: To make soymilk ice cubes, pour soymilk into plastic ice cube trays (preferably the easy-to-release variety). Freeze until completely firm; then unmold the cubes onto a cold pan or plate. Transfer the cubes to a re-sealable freezer bag and store them in the freezer. Use as needed.

Açai, Blackberry, Raspberry, Banana, Vanilla Soymilk, Blackberry Soy Yogurt, and Granola

1 SERVING

1 cup (or more) açai juice

1 tablespoon (or to taste) honey

1 cup almost frozen blackberries

½ cup almost frozen raspberries

½ cup almost frozen diced banana

2 tablespoons frozen vanilla soymilk cubes (see note)

1 (6-ounce) container blackberry (or favorite flavor) soy yogurt

2 to 4 tablespoons favorite granola

Combine the açai juice, honey, blackberries, raspberries, banana, vanilla soymilk cubes, and blackberry soy yogurt in a blender container in the order listed.

Cover the container; and then turn on the blender. Press the pulse button on its lowest blade-speed setting; and process until the ingredients are mostly blended.

Continue to mix on the highest blade-speed setting button until the mixture is smooth (it may be necessary to turn off the blender periodically to stir the mixture with a spoon, always working from the bottom up).

Turn off the blender. Add the granola and mix until just blended. Scrape the smoothie into a glass.

Note: To make soymilk ice cubes, pour soymilk into plastic ice cube trays (preferably the easy-to-release variety). Freeze until completely firm; then unmold the cubes onto a cold pan or plate. Transfer the cubes to a re-sealable freezer bag and store them in the freezer. Use as needed.

Açai, Blueberry, Apricot, Mango, and Apricot Soy Yogurt

1 SERVING

1 cup (or more) açai juice

1 tablespoon (or to taste) honey

1 cup almost frozen blueberries

½ cup almost frozen diced apricots

½ cup almost frozen diced mango

1 (6-ounce) container apricot (or favorite flavor) soy yogurt

Combine all the ingredients in a blender container in the order listed.

Cover the container; and then turn on the blender. Press the pulse button on its lowest blade-speed setting; and process until the ingredients are mostly blended.

Continue to mix on the highest blade-speed setting button until the mixture is smooth (it may be necessary to turn off the blender periodically to stir the mixture with a spoon, always working from the bottom up).

Turn off the blender. Scrape the smoothie into a glass.

Açai, Blueberry, Apricot, Peach, Tofu, Vanilla Soymilk, and Vanilla Soy Yogurt

1 SERVING

1 cup (or more) açai juice

1 tablespoon (or to taste) honey

1 cup almost frozen blueberries

½ cup almost frozen diced apricots

½ cup almost frozen diced peach

½ cup firm tofu, cut into large chunks

2 tablespoons frozen vanilla soymilk cubes (see note)

1 (6-ounce) container vanilla (or favorite flavor) soy yogurt

Combine all the ingredients in a blender container in the order listed.

Cover the container; and then turn on the blender. Press the pulse button on its lowest blade-speed setting; and process until the ingredients are mostly blended.

Continue to mix on the highest blade-speed setting button until the mixture is smooth (it may be necessary to turn off the blender periodically to stir the mixture with a spoon, always working from the bottom up).

Turn off the blender. Scrape the smoothie into a glass.

Note: To make soymilk ice cubes, pour soymilk into plastic ice cube trays (preferably the easy-to-release variety). Freeze until completely firm; then unmold the cubes onto a cold pan or plate. Transfer the cubes to a re-sealable freezer bag and store them in the freezer. Use as needed.

Açai, Blueberry, Blackberry, Vanilla Soymilk, Blueberry Soy Yogurt, and Protein Powder

1 TO 2 SERVINGS

1 cup (or more) açai juice

1 tablespoon (or to taste) honey

1 cup almost frozen blueberries

1 cup almost frozen blackberries

½ cup frozen vanilla soymilk cubes (see note)

1 (6-ounce) container blueberry (or favorite flavor) soy yogurt

1 to 2 tablespoons favorite vanilla protein powder

Combine all the ingredients in a blender container in the order listed.

Cover the container; and then turn on the blender. Press the pulse button on its lowest blade-speed setting; and process until the ingredients are mostly blended.

Continue to mix on the highest blade-speed setting button until the mixture is smooth (it may be necessary to turn off the blender periodically to stir the mixture with a spoon, always working from the bottom up).

Turn off the blender. Scrape the smoothie into a glass.

Note: To make soymilk ice cubes, pour soymilk into plastic ice cube trays (preferably the easy-to-release variety). Freeze until completely firm; then unmold the cubes onto a cold pan or plate. Transfer the cubes to a re-sealable freezer bag and store them in the freezer. Use as needed.

Açai, Blueberry, Cherry, Banana, Tofu, and Vanilla Soymilk

1 SERVING

1 cup (or more) açai juice
1 tablespoon (or to taste) honey
1 cup almost frozen blueberries
½ cup almost frozen cherries
½ cup almost frozen diced banana
½ cup firm tofu, cut into chunks
¼ cup frozen vanilla soy milk cubes (see note)

Combine all the ingredients in a blender container in the order listed.

Cover the container; and then turn on the blender. Press the pulse button on its lowest blade-speed setting; and process until the ingredients are mostly blended.

Continue to mix on the highest blade-speed setting button until the mixture is smooth (it may be necessary to turn off the blender periodically to stir the mixture with a spoon, always working from the bottom up).

Turn off the blender. Scrape the smoothie into a glass.

Note: To make soymilk ice cubes, pour soymilk into plastic ice cube trays (preferably the easy-to-release variety). Freeze until completely firm; then unmold the cubes onto a cold pan or plate. Transfer the cubes to a re-sealable freezer bag and store them in the freezer. Use as needed.

Açai, Blueberry, Cherry, Raspberry, Vanilla Soymilk, and Cherry Soy Yogurt

1 SERVING

1 cup (or more) açai juice

1 tablespoon (or to taste) honey

1 cup almost frozen blueberries

½ cup almost frozen cherries

½ cup almost frozen raspberries

¼ cup frozen vanilla soymilk cubes (see note)

1 (6-ounce) container cherry (or favorite flavor)
soy yogurt

Combine all the ingredients in a blender container in the order listed.

Cover the container; and then turn on the blender. Press the pulse button on its lowest blade-speed setting; and process until the ingredients are mostly blended.

Continue to mix on the highest blade-speed setting button until the mixture is smooth (it may be necessary to turn off the blender periodically to stir the mixture with a spoon, always working from the bottom up).

Turn off the blender. Scrape the smoothie into a glass.

Note: To make soymilk ice cubes, pour soymilk into plastic ice cube trays (preferably the easy-to-release variety). Freeze until completely firm; then unmold the cubes onto a cold pan or plate. Transfer the cubes to a re-sealable freezer bag and store them in the freezer. Use as needed.

Açai, Blueberry, Pineapple, Peach, Tofu, and Peach Soy Yogurt

1 SERVING

1 cup (or more) açai juice

1 tablespoon (or to taste) honey

1 cup almost frozen blueberries

½ cup almost frozen diced pineapple

½ cup almost frozen diced peach

½ cup firm tofu, cut into chunks

1 (6-ounce) container peach (or favorite flavor) soy yogurt

Combine all the ingredients in a blender container in the order listed.

Cover the container; and then turn on the blender. Press the pulse button on its lowest blade-speed setting; and process until the ingredients are mostly blended.

Continue to mix on the highest blade-speed setting button until the mixture is smooth (it may be necessary to turn off the blender periodically to stir the mixture with a spoon, always working from the bottom up).

Turn off the blender. Scrape the smoothie into a glass.

Açai, Blueberry, Raspberry, Kiwi, Pomegranate, Raspberry Soy Yogurt, and Protein Powder

1 TO 2 SERVINGS

1 cup (or more) açai juice

1 tablespoon (or to taste) honey

1 cup almost frozen blueberries

½ cup almost frozen raspberries

½ cup almost frozen diced kiwi

½ cup frozen pomegranate juice cubes (see note)

1 (6-ounce) container raspberry (or favorite flavor) soy yogurt

1 to 2 tablespoons favorite vanilla protein powder

Combine all the ingredients in a blender container in the order listed.

Cover the container; and then turn on the blender. Press the pulse button on its lowest blade-speed setting; and process until the ingredients are mostly blended.

Continue to mix on the highest blade-speed setting button until the mixture is smooth (it may be necessary to turn off the blender periodically to stir the mixture with a spoon, always working from the bottom up).

Turn off the blender. Scrape the smoothie into a glass.

Note: To make pomegranate ice cubes, pour pomegranate juice into plastic ice cube trays (preferably the easy-to-release variety). Freeze until completely firm; then unmold the cubes onto a cold pan or plate. Transfer the cubes to a re-sealable freezer bag and store them in the freezer. Use as needed.

Açai, Blueberry, Raspberry, Vanilla Soymilk, and Wheat Germ

1 TO 2 SERVINGS

1 cup (or more) açai juice
1 tablespoon (or to taste) honey
1 cup almost frozen blueberries
1 cup almost frozen raspberries
½ cup frozen vanilla soymilk cubes (see note)
2 to 4 tablespoons wheat germ (see note)

Combine the açai juice, honey, blueberries, raspberries, and vanilla soymilk cubes in the order listed.

Cover the container; and then turn on the blender. Press the pulse button on its lowest blade-speed setting; and process until the ingredients are mostly blended.

Continue to mix on the highest blade-speed setting button until the mixture is smooth (it may be necessary to turn off the blender periodically to stir the mixture with a spoon, always working from the bottom up).

Turn off the blender. Add the wheat germ and mix until just blended. Scrape the smoothie into a glass.

Note: To make soymilk ice cubes, pour soymilk into plastic ice cube trays (preferably the easy-to-release variety). FFreeze until completely firm; then unmold the cubes onto a cold pan or plate. Transfer the cubes to a re-sealable freezer bag and store them in the freezer. Use as needed.

Note: To toast wheat germ, spread ¼ cup (or more) wheat germ in a single layer on a rimmed cookie sheet. Bake in a preheated 350 degrees F oven for about 5 minutes.

Açai, Blueberry, Strawberry, Banana, Blueberry Soy Yogurt, and Granola

1 SERVING

1 cup (or more) açai juice

1 tablespoon (or to taste) honey

1 cup almost frozen blueberries

½ cup almost frozen diced strawberries

½ cup almost frozen diced banana

1 (6-ounce) container blueberry (or favorite flavor) soy yogurt

2 to 4 tablespoons favorite granola

Combine the açai juice, honey, blueberries, strawberries, banana, and blueberry soy yogurt in a blender container in the order listed.

Cover the container; and then turn on the blender. Press the pulse button on its lowest blade-speed setting; and process until the ingredients are mostly blended.

Continue to mix on the highest blade-speed setting button until the mixture is smooth (it may be necessary to turn off the blender periodically to stir the mixture with a spoon, always working from the bottom up).

Turn off the blender. Add the granola and mix until just blended. Scrape the smoothie into a glass.

Açai, Blueberry, Strawberry, Vanilla Soymilk, and Vanilla Soy Yogurt

1 SERVING

1 cup (or more) açai juice
1 tablespoon (or to taste) honey
1 cup almost frozen blueberries
1 cup almost frozen diced strawberries
½ cup frozen vanilla soymilk cubes (see note)
1 (6-ounce) container vanilla (or favorite flavor)
soy yogurt

Combine all the ingredients in a blender container in the order listed.

Cover the container; and then turn on the blender. Press the pulse button on its lowest blade-speed setting; and process until the ingredients are mostly blended.

Continue to mix on the highest blade-speed setting button until the mixture is smooth (it may be necessary to turn off the blender periodically to stir the mixture with a spoon, always working from the bottom up).

Turn off the blender. Scrape the smoothie into a glass.

Note: To make soymilk ice cubes, pour soymilk into plastic ice cube trays (preferably the easy-to-release variety). Freeze until completely firm; then unmold the cubes onto a cold pan or plate. Transfer the cubes to a re-sealable freezer bag and store them in the freezer. Use as needed.

Açai, Cantaloupe, Cherry, Banana, Cherry Soy Yogurt, and Wheat Germ

1 SERVING

1 cup (or more) açai juice

1 tablespoon (or to taste) honey

1 cup almost frozen diced cantaloupe

½ cup almost frozen cherries

½ cup almost frozen diced banana

1 (6-ounce) container cherry (or favorite flavor) soy yogurt

2 to 4 tablespoons wheat germ (see note)

Combine the açai juice, honey, cantaloupe, cherries, banana, and cherry soy yogurt in a blender container in the order listed.

Cover the container; and then turn on the blender. Press the pulse button on its lowest blade-speed setting; and process until the ingredients are mostly blended.

Continue to mix on the highest blade-speed setting button until the mixture is smooth (it may be necessary to turn off the blender periodically to stir the mixture with a spoon, always working from the bottom up).

Turn off the blender. Add the wheat germ and mix until just blended. Scrape the smoothie into a glass.

Note: To toast wheat germ, spread ¼ cup (or more) wheat germ in a single layer on a rimmed cookie sheet. Bake in a preheated 350 degrees F oven for about 5 minutes.

Açai, Cherry, Apricot, Vanilla Soymilk, Cherry Soy Yogurt, and Protein Powder

1 TO 2 SERVINGS

1 cup (or more) açai juice

1 tablespoon (or to taste) honey

1 cup almost frozen cherries

1 cup almost frozen diced apricots

¼ cup frozen vanilla soymilk cubes (see note)

1 (6-ounce) container cherry (or favorite flavor) soy yogurt

1 to 2 tablespoons favorite vanilla protein powder

Combine all the ingredients in a blender container in the order listed.

Cover the container; and then turn on the blender. Press the pulse button on its lowest blade-speed setting; and process until the ingredients are mostly blended.

Continue to mix on the highest blade-speed setting button until the mixture is smooth (it may be necessary to turn off the blender periodically to stir the mixture with a spoon, always working from the bottom up).

Turn off the blender. Scrape the smoothie into a glass.

Note: To make soymilk ice cubes, pour soymilk into plastic ice cube trays (preferably the easy-to-release variety). Freeze until completely firm; then unmold the cubes onto a cold pan or plate. Transfer the cubes to a re-sealable freezer bag and store them in the freezer. Use as needed.

Açai, Cherry, Blueberry, Banana, Tofu, and Blueberry Soy Yogurt

1 SERVING

1 cup (or more) açai juice
1 tablespoon (or to taste) honey
1 cup almost frozen cherries
½ cup almost frozen blueberries
½ cup almost frozen diced banana
½ cup firm tofu, cut into chunks
1 (6-ounce) container blueberry (or favorite flavor) soy yogurt

Combine all the ingredients in a blender container in the order listed.

Cover the container; and then turn on the blender. Press the pulse button on its lowest blade-speed setting; and process until the ingredients are mostly blended.

Continue to mix on the highest blade-speed setting button until the mixture is smooth (it may be necessary to turn off the blender periodically to stir the mixture with a spoon, always working from the bottom up).

Turn off the blender. Scrape the smoothie into a glass.

Açai, Cherry, Pineapple, Banana, Tofu, Vanilla Soymilk, and Wheat Germ

❧

1 SERVING

1 cup (or more) açai juice

1 tablespoon (or to taste) honey

1 cup almost frozen cherries

½ cup almost frozen diced pineapple

½ cup almost frozen diced banana

½ cup firm tofu, cut up in large chunks

2 tablespoons frozen vanilla soymilk cubes (see note)

2 to 4 tablespoons wheat germ (see note)

Combine the açai juice, honey, cherries, pineapple, banana, vanilla soymilk cubes, and tofu in a blender container in the order listed.

Cover the container; and then turn on the blender. Press the pulse button on its lowest blade-speed setting; and process until the ingredients are mostly blended.

Continue to mix on the highest blade-speed setting button until the mixture is smooth (it may be necessary to turn off the blender periodically to stir the mixture with a spoon, always working from the bottom up).

Turn off the blender. Add the wheat germ and mix until just blended. Scrape the smoothie into a glass.

Note: To make soymilk ice cubes, pour soymilk into plastic ice cube trays (preferably the easy-to-release variety). Freeze until completely firm; then unmold the cubes onto a cold pan or plate. Transfer the cubes to a re-sealable freezer bag and store them in the freezer. Use as needed.

Note: To toast wheat germ, spread ¼ cup (or more) wheat germ in a single layer on a rimmed cookie sheet. Bake in a preheated 350 degrees F oven for about 5 minutes.

Açai, Cherry, Raspberry, Banana, Tofu, and Raspberry Soy Yogurt

1 SERVING

1 cup (or more) açai juice
1 tablespoon (or to taste) honey
1 cup almost frozen cherries
½ cup almost frozen raspberries
½ cup almost frozen diced banana
½ cup firm tofu, cut into chunks
1 (6-ounce) container raspberry (or favorite flavor) soy yogurt

Combine all the ingredients in a blender container in the order listed.

Cover the container; and then turn on the blender. Press the pulse button on its lowest blade-speed setting; and process until the ingredients are mostly blended.

Continue to mix on the highest blade-speed setting button until the mixture is smooth (it may be necessary to turn off the blender periodically to stir the mixture with a spoon, always working from the bottom up).

Turn off the blender. Scrape the smoothie into a glass.

Açai, Flaxseed, Banana, Raspberry, Vanilla Soymilk, and Raspberry Soy Yogurt

1 SERVING

1 cup (or more) açai juice

1 tablespoon (or to taste) honey

1 tablespoon flaxseed oil (or according to manufacturer's recommendation)

1 cup almost frozen diced banana

1 cup almost frozen raspberries

¼ cup frozen vanilla soymilk cubes (see note)

1 (6-ounce) container raspberry (or favorite flavor) soy yogurt

Combine all the ingredients in a blender container in the order listed.

Cover the container; and then turn on the blender. Press the pulse button on its lowest blade-speed setting; and process until the ingredients are mostly blended.

Continue to mix on the highest blade-speed setting button until the mixture is smooth (it may be necessary to turn off the blender periodically to stir the mixture with a spoon, always working from the bottom up).

Turn off the blender. Scrape the smoothie into a glass.

Note: To make soymilk ice cubes, pour soymilk into plastic ice cube trays (preferably the easy-to-release variety). Freeze until completely firm; then unmold the cubes onto a cold pan or plate. Transfer the cubes to a re-sealable freezer bag and store them in the freezer. Use as needed.

Açai, Flaxseed, Blueberry, Apricot, Mango, Pomegranate, and Blueberry Soy Yogurt

1 SERVING

1 cup (or more) açai uice

1 tablespoon (or to taste) honey

1 tablespoon flaxseed oil (or according to manufacturer's recommendation)

1 cup almost frozen blueberries

½ cup almost frozen diced apricots

½ cup almost frozen diced mango

¼ cup frozen pomegranate juice cubes (see note)

1 (6-ounce) container blueberry (or favorite flavor) soy yogurt

2 to 4 tablespoons favorite granola, optional

Combine the açai juice, honey, flaxseed oil, blueberries, apricots, mango, pomegranate juice cubes, and blueberry soy yogurt in a blender container in the order listed.

Cover the container; and then turn on the blender. Press the pulse button on its lowest blade-speed setting; and process until the ingredients are mostly blended.

Continue to mix on the highest blade-speed setting button until the mixture is smooth (it may be necessary to turn off the blender periodically to stir the mixture with a spoon, always working from the bottom up).

Turn off the blender. Add the optional granola and mix until just blended. Scrape the smoothie into a glass.

Note: To make pomegranate ice cubes, pour pomegranate juice into plastic ice cube trays (preferably the easy-to-release variety). Freeze until completely firm; then unmold the cubes onto a cold pan or plate. Transfer the cubes to a re-sealable freezer bag and store them in the freezer. Use as needed.

Açai, Flaxseed, Blueberry, Cantaloupe, Peach, Pomegranate, and Blueberry Soy Yogurt

1 SERVING

1 cup (or more) açai juice

1 tablespoon (or to taste) honey

1 tablespoon flaxseed oil (or according to manufacturer's recommendation)

1 cup almost frozen blueberries

½ cup almost frozen diced cantaloupe

½ cup almost frozen diced peach

¼ cup frozen pomegranate juice cubes (see note)

1 (6-ounce) container blueberry (or favorite flavor) soy yogurt

Combine all the ingredients in a blender container in the order listed.

Cover the container; and then turn on the blender. Press the pulse button on its lowest blade-speed setting; and process until the ingredients are mostly blended.

Continue to mix on the highest blade-speed setting button until the mixture is smooth (it may be necessary to turn off the blender periodically to stir the mixture with a spoon, always working from the bottom up).

Turn off the blender. Scrape the smoothie into a glass.

Note: To make pomegranate ice cubes, pour pomegranate juice into plastic ice cube trays (preferably the easy-to-release variety). Freeze until completely firm; then unmold the cubes onto a cold pan or plate. Transfer the cubes to a re-sealable freezer bag and store them in the freezer. Use as needed.

Açai, Flaxseed, Blueberry, Strawberry, Banana, and Strawberry Soy Yogurt

1 SERVING

1 cup (or more) açai juice

1 tablespoon (or to taste) honey

1 tablespoon flaxseed oil (or according to manufacturer's recommendation)

1 cup almost frozen blueberries

½ cup almost frozen diced strawberries

½ cup almost frozen diced banana

1 (6-ounce) container strawberry (or favorite flavor) soy yogurt

Combine all the ingredients in a blender container in the order listed.

Cover the container; and then turn on the blender. Press the pulse button on its lowest blade-speed setting; and process until the ingredients are mostly blended.

Continue to mix on the highest blade-speed setting button until the mixture is smooth (it may be necessary to turn off the blender periodically to stir the mixture with a spoon, always working from the bottom up).

Turn off the blender. Scrape the smoothie into a glass.

Açai, Flaxseed, Mango, Blueberry, Raspberry, Vanilla Soymilk, and Raspberry Soy Yogurt

1 SERVING

1 cup (or more) açai juice
1 tablespoon (or to taste) honey
1 tablespoon flaxseed oil (or according to manufacturer's recommendation)
½ cup almost frozen diced mango
½ cup almost frozen blueberries
½ cup almost frozen raspberries
½ cup frozen vanilla soymilk cubes (see note)
1 (6-ounce) container raspberry (or favorite flavor) soy yogurt

Combine all the ingredients in a blender container in the order listed.

Cover the container; and then turn on the blender. Press the pulse button on its lowest blade-speed setting; and process until the ingredients are mostly blended.

Continue to mix on the highest blade-speed setting button until the mixture is smooth (it may be necessary to turn off the blender periodically to stir the mixture with a spoon, always working from the bottom up).

Turn off the blender. Scrape the smoothie into a glass.

Note: To make soymilk ice cubes, pour soymilk into plastic ice cube trays (preferably the easy-to-release variety). Freeze until completely firm; then unmold the cubes onto a cold pan or plate. Transfer the cubes to a re-sealable freezer bag and store them in the freezer. Use as needed.

Açai, Flaxseed, Peach, Pineapple, Kiwi, Vanilla Soymilk, and Peach Soy Yogurt

1 SERVING

- 1 cup (or more) açai juice
- 1 tablespoon (or to taste) honey
- 1 tablespoon flaxseed oil (or according to manufacturer's recommendation)
- 1 cup almost frozen diced peach
- ½ cup almost frozen diced pineapple
- ½ cup almost frozen diced kiwi
- ¼ cup frozen vanilla soymilk cubes (see note)
- 1 (6-ounce) container peach (or favorite flavor) soy yogurt

Combine all the ingredients in a blender container in the order listed.

Cover the container; and then turn on the blender. Press the pulse button on its lowest blade-speed setting; and process until the ingredients are mostly blended.

Continue to mix on the highest blade-speed setting button until the mixture is smooth (it may be necessary to turn off the blender periodically to stir the mixture with a spoon, always working from the bottom up).

Turn off the blender. Scrape the smoothie into a glass.

Note: To make soymilk ice cubes, pour soymilk into plastic ice cube trays (preferably the easy-to-release variety). Freeze until completely firm; then unmold the cubes onto a cold pan or plate. Transfer the cubes to a re-sealable freezer bag and store them in the freezer. Use as needed.

Açai, Flaxseed, Peach, Raspberry, Pomegranate, and Raspberry Soy Yogurt

1 SERVING

1 cup (or more) açai juice

1 tablespoon (or to taste) honey

1 tablespoon flaxseed oil (or according to manufacturer's recommendation)

1 cup almost frozen diced peach

1 cup almost frozen raspberries

½ cup frozen pomegranate juice cubes (see note)

1 (6-ounce) container raspberry (or favorite flavor) soy yogurt

Combine all the ingredients in a blender container in the order listed.

Cover the container; and then turn on the blender. Press the pulse button on its lowest blade-speed setting; and process until the ingredients are mostly blended.

Continue to mix on the highest blade-speed setting button until the mixture is smooth (it may be necessary to turn off the blender periodically to stir the mixture with a spoon, always working from the bottom up).

Turn off the blender. Scrape the smoothie into a glass.

Note: To make pomegranate ice cubes, pour pomegranate juice into plastic ice cube trays (preferably the easy-to-release variety). Freeze until completely firm; then unmold the cubes onto a cold pan or plate. Transfer the cubes to a re-sealable freezer bag and store them in the freezer. Use as needed.

Açai, Flaxseed, Pineapple, Peach, Banana, and Peach Soy Yogurt

1 SERVING

1 cup (or more) açai juice

1 tablespoon (or to taste) honey

1 tablespoon flaxseed oil (or according to manufacturer's recommendation)

1 cup almost frozen diced pineapple

½ cup almost frozen diced peach

½ cup almost frozen diced banana

1 (6-ounce) container peach (or favorite flavor) soy yogurt

Combine all the ingredients in a blender container in the order listed.

Cover the container; and then turn on the blender. Press the pulse button on its lowest blade-speed setting; and process until the ingredients are mostly blended.

Continue to mix on the highest blade-speed setting button until the mixture is smooth (it may be necessary to turn off the blender periodically to stir the mixture with a spoon, always working from the bottom up).

Turn off the blender. Scrape the smoothie into a glass.

Açai, Flaxseed, Raspberry, Pineapple, Blueberry, and Vanilla Soy Yogurt

1 SERVING

1 cup (or more) açai juice

1 tablespoon (or to taste) honey

1 tablespoon flaxseed oil (or according to manufacturer's recommendation)

1 cup almost frozen raspberries

½ cup almost frozen diced pineapple

½ cup almost frozen blueberries

1 (6-ounce) container vanilla (or favorite flavor) soy yogurt

Combine all the ingredients in a blender container in the order listed.

Cover the container; and then turn on the blender. Press the pulse button on its lowest blade-speed setting; and process until the ingredients are mostly blended.

Continue to mix on the highest blade-speed setting button until the mixture is smooth (it may be necessary to turn off the blender periodically to stir the mixture with a spoon, always working from the bottom up).

Turn off the blender. Scrape the smoothie into a glass.

Açai, Mango, Cantaloupe, Pineapple, Vanilla Soymilk, Mango Soy Yogurt, and Granola

1 SERVING

1 cup (or more) açai juice
1 tablespoon (or to taste) honey
1 cup almost frozen diced mango
½ cup almost frozen diced cantaloupe
½ cup almost frozen diced pineapple
½ cup frozen vanilla soymilk cubes (see note)
1 (6-ounce) container mango (or favorite flavor) soy yogurt
2 to 4 tablespoons favorite granola

Combine the açai juice, honey, mango, cantaloupe, pineapple, vanilla soymilk cubes, and mango soy yogurt in a blender container in the order listed.

Cover the container; and then turn on the blender. Press the pulse button on its lowest blade-speed setting; and process until the ingredients are mostly blended.

Continue to mix on the highest blade-speed setting button until the mixture is smooth (it may be necessary to turn off the blender periodically to stir the mixture with a spoon, always working from the bottom up).

Turn off the blender. Add the granola and mix until just blended. Scrape the smoothie into a glass.

Note: To make soymilk ice cubes, pour soymilk into plastic ice cube trays (preferably the easy-to-release variety). Freeze until completely firm; then unmold the cubes onto a cold pan or plate. Transfer the cubes to a re-sealable freezer bag and store them in the freezer. Use as needed.

Açai, Mango, Cherry, Pineapple, Vanilla Soymilk, Mango Soy Yogurt, and Wheat Germ

1 SERVING

1 cup (or more) açai juice

1 tablespoon (or to taste) honey

1 cup almost frozen diced mango

½ cup almost frozen cherries

½ cup almost frozen diced pineapple

¼ cup frozen vanilla soymilk cubes (see note)

1 (6-ounce) container mango (or favorite flavor) soy yogurt

2 to 4 tablespoons wheat germ (see note)

Combine the açai juice, honey, mango, cherries, pineapple, vanilla soymilk cubes, and mango soy yogurt in a blender container in the order listed.

Cover the container; and then turn on the blender. Press the pulse button on its lowest blade-speed setting; and process until the ingredients are mostly blended.

Continue to mix on the highest blade-speed setting button until the mixture is smooth (it may be necessary to turn off the blender periodically to stir the mixture with a spoon, always working from the bottom up).

Turn off the blender. Add the wheat germ and mix until just blended. Scrape the smoothie into a glass.

Note: To make soymilk ice cubes, pour soymilk into plastic ice cube trays (preferably the easy-to-release variety). Freeze until completely firm; then unmold the cubes onto a cold pan or plate. Transfer the cubes to a re-sealable freezer bag and store them in the freezer. Use as needed.

Note: To toast wheat germ, spread ¼ cup (or more) wheat germ in a single layer on a rimmed cookie sheet. Bake in a preheated 350 degrees F oven for about 5 minutes.

Açai, Mango, Pineapple, Peach, and Peach Soy Yogurt

1 SERVING

1 cup (or more) açai juice
1 tablespoon (or to taste) honey
1 cup almost frozen diced mango
½ cup almost frozen diced pineapple
½ cup almost frozen diced peach
1 (6-ounce) container peach (or favorite flavor) soy yogurt

Combine all the ingredients in a blender container in the order listed.

Cover the container; and then turn on the blender. Press the pulse button on its lowest blade-speed setting; and process until the ingredients are mostly blended.

Continue to mix on the highest blade-speed setting button until the mixture is smooth (it may be necessary to turn off the blender periodically to stir the mixture with a spoon, always working from the bottom up).

Turn off the blender. Scrape the smoothie into a glass.

Açai, Peach, Mango, Banana, Vanilla Soymilk, Peach Soy Yogurt, and Granola

1 SERVING

1 cup (or more) açai juice

1 tablespoon (or to taste) honey

1 cup almost frozen diced peach

½ cup almost frozen diced mango

½ cup almost frozen diced banana

2 tablespoons frozen vanilla soymilk cubes
(see note)

1 (6-ounce) container peach (or favorite flavor)
soy yogurt

2 to 4 tablespoons favorite granola

Combine the açai juice, honey, peach, mango, banana, vanilla soymilk cubes, and peach soy yogurt in a blender container in the order listed.

Cover the container; and then turn on the blender. Press the pulse button on its lowest blade-speed setting; and process until the ingredients are mostly blended.

Continue to mix on the highest blade-speed setting button until the mixture is smooth (it may be necessary to turn off the blender periodically to stir the mixture with a spoon, always working from the bottom up).

Turn off the blender. Add the granola and mix until just blended. Scrape the smoothie into a glass.

Note: To make soymilk ice cubes, pour soymilk into plastic ice cube trays (preferably the easy-to-release variety). Freeze until completely firm; then unmold the cubes onto a cold pan or plate. Transfer the cubes to a re-sealable freezer bag and store them in the freezer. Use as needed.

Açai, Peach, Mango, Orange, Peach Soy Yogurt, and Protein Powder

1 TO 2 SERVINGS

1 cup (or more) açai juice

1 tablespoon (or to taste) honey

1 cup almost frozen diced peach

1 cup almost frozen diced mango

¼ cup frozen orange juice cubes (see note)

1 (6-ounce) container peach (or favorite flavor) soy yogurt

1 to 2 tablespoons favorite vanilla protein powder

Combine all the ingredients in a blender container in the order listed.

Cover the container; and then turn on the blender. Press the pulse button on its lowest blade-speed setting; and process until the ingredients are mostly blended.

Continue to mix on the highest blade-speed setting button until the mixture is smooth (it may be necessary to turn off the blender periodically to stir the mixture with a spoon, always working from the bottom up).

Turn off the blender. Scrape the smoothie into a glass.

Note: To make orange ice cubes, pour orange juice into plastic ice cube trays (preferably the easy-to-release variety). Freeze until completely firm; then unmold the cubes onto a cold pan or plate. Transfer the cubes to a re-sealable freezer bag and store them in the freezer. Use as needed.

Açai, Pineapple, Apricot, Mango, Vanilla Soymilk, and Apricot Mango Soy Yogurt

1 SERVING

1 cup (or more) açai juice

1 tablespoon (or to taste) honey

1 cup almost frozen diced pineapple

½ cup almost frozen diced apricots

½ cup almost frozen diced mango

¼ cup frozen vanilla soymilk cubes (see note)

1 (6-ounce) container apricot mango (or favorite flavor) soy yogurt

Combine all the ingredients in a blender container in the order listed.

Cover the container; and then turn on the blender. Press the pulse button on its lowest blade-speed setting; and process until the ingredients are mostly blended.

Continue to mix on the highest blade-speed setting button until the mixture is smooth (it may be necessary to turn off the blender periodically to stir the mixture with a spoon, always working from the bottom up).

Turn off the blender. Scrape the smoothie into a glass.

Note: To make soymilk ice cubes, pour soymilk into plastic ice cube trays (preferably the easy-to-release variety). Freeze until completely firm; then unmold the cubes onto a cold pan or plate. Transfer the cubes to a re-sealable freezer bag and store them in the freezer. Use as needed.

Açai, Pineapple, Banana, Peach, Vanilla Soymilk, Peach Soy Yogurt, and Granola

1 SERVING

1 cup (or more) açai juice

1 tablespoon (or to taste) honey

1 cup almost frozen diced pineapple

½ cup almost frozen diced banana

½ cup almost frozen diced peach

¼ cup frozen vanilla soymilk cubes (see note)

1 (6-ounce) container peach (or favorite flavor) soy yogurt

2 to 4 tablespoons granola

Combine the açai juice, honey, pineapple, banana, peach, vanilla soymilk cubes, and peach soy yogurt in a blender container in the order listed.

Cover the container; and then turn on the blender. Press the pulse button on its lowest blade-speed setting; and process until the ingredients are mostly blended.

Continue to mix on the highest blade-speed setting button until the mixture is smooth (it may be necessary to turn off the blender periodically to stir the mixture with a spoon, always working from the bottom up).

Turn off the blender. Add the granola and mix until just blended. Scrape the smoothie into a glass.

Note: To make soymilk ice cubes, pour soymilk into plastic ice cube trays (preferably the easy-to-release variety). Freeze until completely firm; then unmold the cubes onto a cold pan or plate. Transfer the cubes to a re-sealable freezer bag and store them in the freezer. Use as needed.

Açai, Pineapple, Blueberry, Tofu, and Vanilla Soymilk

1 SERVING

1 cup (or more) açai juice
1 tablespoon (or to taste) honey
1 cup almost frozen diced pineapple
1 cup almost frozen blueberries
½ cup firm tofu, cut into large chunks
½ cup frozen vanilla soymilk cubes (see note)

Combine all the ingredients in a blender container in the order listed.

Cover the container; and then turn on the blender. Press the pulse button on its lowest blade-speed setting; and process until the ingredients are mostly blended.

Continue to mix on the highest blade-speed setting button until the mixture is smooth (it may be necessary to turn off the blender periodically to stir the mixture with a spoon, always working from the bottom up).

Turn off the blender. Scrape the smoothie into a glass.

Note: To make soymilk ice cubes, pour soymilk into plastic ice cube trays (preferably the easy-to-release variety). Freeze until completely firm; then unmold the cubes onto a cold pan or plate. Transfer the cubes to a re-sealable freezer bag and store them in the freezer. Use as needed.

Açai, Pineapple, Mango, Pomegranate, and Vanilla Soy Yogurt

❧❧

1 SERVING

½ cup (or more) açai juice

1 tablespoon (or to taste) honey

1 cup almost frozen diced pineapple

1 cup almost frozen diced mango

½ cup frozen pomegranate juice cubes (see note)

1 (6-ounce) container vanilla (or favorite flavor) soy yogurt

Combine all the ingredients in a blender container in the order listed.

Cover the container; and then turn on the blender. Press the pulse button on its lowest blade-speed setting; and process until the ingredients are mostly blended.

Continue to mix on the highest blade-speed setting button until the mixture is smooth (it may be necessary to turn off the blender periodically to stir the mixture with a spoon, always working from the bottom up).

Turn off the blender. Scrape the smoothie into a glass.

Note: To make pomegranate ice cubes, pour pomegranate juice into plastic ice cube trays (preferably the easy-to-release variety). Freeze until completely firm; then unmold the cubes onto a cold pan or plate. Transfer the cubes to a re-sealable freezer bag and store them in the freezer. Use as needed.

Açai, Pineapple, Peach, Banana, Tofu, and Peach Soy Yogurt

1 SERVING

1 cup (or more) açai juice
1 tablespoon (or to taste) honey
1 cup almost frozen diced pineapple
½ cup almost frozen diced peach
½ cup almost frozen diced banana
½ cup firm tofu, cut up in chunks
1 (6-ounce) container peach (or favorite flavor)
   soy yogurt

Combine all the ingredients in a blender container in the order listed.

Cover the container; and then turn on the blender. Press the pulse button on its lowest blade-speed setting; and process until the ingredients are mostly blended.

Continue to mix on the highest blade-speed setting button until the mixture is smooth (it may be necessary to turn off the blender periodically to stir the mixture with a spoon, always working from the bottom up).

Turn off the blender. Scrape the smoothie into a glass.

Açai, Pineapple, Peach, Tofu, Vanilla Soymilk, and Peach Soy Yogurt

❧❧❧

1 SERVING

1 cup (or more) açai juice
1 tablespoon (or to taste) honey
1 cup almost frozen diced pineapple
1 cup almost frozen diced peach
½ cup firm tofu, cut into chunks
¼ cup frozen vanilla soymilk cubes (see note)
1 (6-ounce) container peach (or favorite flavor) soy yogurt

Combine all the ingredients in a blender container in the order listed.

Cover the container; and then turn on the blender. Press the pulse button on its lowest blade-speed setting; and process until the ingredients are mostly blended.

Continue to mix on the highest blade-speed setting button until the mixture is smooth (it may be necessary to turn off the blender periodically to stir the mixture with a spoon, always working from the bottom up).

Turn off the blender. Scrape the smoothie into a glass.

Note: To make soymilk ice cubes, pour soymilk into plastic ice cube trays (preferably the easy-to-release variety). Freeze until completely firm; then unmold the cubes onto a cold pan or plate. Transfer the cubes to a re-sealable freezer bag and store them in the freezer. Use as needed.

Açai, Pineapple, Raspberry, Banana, Raspberry Soy Yogurt, and Wheat Germ

1 SERVING

1 cup (or more) açai juice

1 tablespoon (or to taste) honey

1 cup almost frozen diced pineapple

½ cup almost frozen raspberries

½ cup almost frozen diced banana

1 (6-ounce) container raspberry (or favorite flavor) soy yogurt

2 to 4 tablespoons wheat germ (see note)

Combine the açai juice, honey, pineapple, raspberries, banana, and raspberry soy yogurt in a blender container in the order listed.

Cover the container; and then turn on the blender. Press the pulse button on its lowest blade-speed setting; and process until the ingredients are mostly blended.

Continue to mix on the highest blade-speed setting button until the mixture is smooth (it may be necessary to turn off the blender periodically to stir the mixture with a spoon, always working from the bottom up).

Turn off the blender. Add the wheat germ and mix until just blended. Scrape the smoothie into a glass.

Note: To toast wheat germ, spread ¼ cup (or more) wheat germ in a single layer on a rimmed cookie sheet. Bake in a preheated 350 degrees F oven for about 5 minutes.

Açai, Pineapple, Strawberry, Pomegranate, Strawberry Soy Yogurt, and Granola

1 SERVING

1 cup (or more) açai juice

1 tablespoon (or to taste) honey

1 cup almost frozen diced pineapple

¾ cup almost frozen diced strawberries

½ cup frozen pomegranate juice cubes (see note)

1 (6-ounce) container strawberry (or favorite flavor) soy yogurt

2 to 4 tablespoons favorite granola

Combine the açai juice, honey, pineapple, strawberries, pomegranate juice cubes, and strawberry soy yogurt in a blender container in the order listed.

Cover the container; and then turn on the blender. Press the pulse button on its lowest blade-speed setting; and process until the ingredients are mostly blended.

Continue to mix on the highest blade-speed setting button until the mixture is smooth (it may be necessary to turn off the blender periodically to stir the mixture with a spoon, always working from the bottom up).

Turn off the blender. Add the granola and mix until just blended. Scrape the smoothie into a glass.

Note: To make pomegranate ice cubes, pour pomegranate juice into plastic ice cube trays (preferably the easy-to-release variety). Freeze until completely firm; then unmold the cubes onto a cold pan or plate. Transfer the cubes to a re-sealable freezer bag and store them in the freezer. Use as needed.

Açai, Raspberry, Cherry, Banana, Pomegranate, and Raspberry Soy Yogurt

1 SERVING

1 cup (or more) açai juice

1 tablespoon (or to taste) honey

1 cup almost frozen raspberries

½ cup almost frozen cherries

½ cup almost frozen diced banana

½ cup frozen pomegranate juice cubes (see note)

1 (6-ounce) container raspberry (or favorite flavor) soy yogurt

Combine all the ingredients in a blender container in the order listed.

Cover the container; and then turn on the blender. Press the pulse button on its lowest blade-speed setting; and process until the ingredients are mostly blended.

Continue to mix on the highest blade-speed setting button until the mixture is smooth (it may be necessary to turn off the blender periodically to stir the mixture with a spoon, always working from the bottom up).

Turn off the blender. Scrape the smoothie into a glass.

Note: To make pomegranate ice cubes, pour pomegranate juice into plastic ice cube trays (preferably the easy-to-release variety). Freeze until completely firm; then unmold the cubes onto a cold pan or plate. Transfer the cubes to a re-sealable freezer bag and store them in the freezer. Use as needed.

Açai, Strawberry, Mango, Pomegranate, Mango Soy Yogurt, and Protein Powder

1 TO 2 SERVINGS

1 cup (or more) açai juice

1 tablespoon (or to taste) honey

1 cup almost frozen diced strawberries

1 cup almost frozen diced mango

¼ cup frozen pomegranate juice cubes (see note)

1 (6-ounce) container mango (or favorite flavor) soy yogurt

1 to 2 tablespoons favorite vanilla protein powder

Combine all the ingredients in a blender container in the order listed.

Cover the container; and then turn on the blender. Press the pulse button on its lowest blade-speed setting; and process until the ingredients are mostly blended.

Continue to mix on the highest blade-speed setting button until the mixture is smooth (it may be necessary to turn off the blender periodically to stir the mixture with a spoon, always working from the bottom up).

Turn off the blender. Scrape the smoothie into a glass.

Note: To make pomegranate ice cubes, pour pomegranate juice into plastic ice cube trays (preferably the easy-to-release variety). Freeze until completely firm; then unmold the cubes onto a cold pan or plate. Transfer the cubes to a re-sealable freezer bag and store them in the freezer. Use as needed.

Açai, Strawberry, Mango, Pomegranate, Strawberry Soy Yogurt, and Wheat Germ

1 SERVING

1 cup (or more) açai juice

1 tablespoon (or to taste) honey

1 cup almost frozen diced strawberries

¾ cup almost frozen diced mango

½ cup frozen pomegranate juice cubes (see note)

1 (6-ounce) container strawberry (or favorite flavor) soy yogurt

2 to 4 tablespoons wheat germ (see note)

Combine the açai juice, honey, strawberries, mango, pomegranate juice cubes, and strawberry soy yogurt in a blender container in the order listed.

Cover the container; and then turn on the blender. Press the pulse button on its lowest blade-speed setting; and process until the ingredients are mostly blended.

Continue to mix on the highest blade-speed setting button until the mixture is smooth (it may be necessary to turn off the blender periodically to stir the mixture with a spoon, always working from the bottom up).

Turn off the blender. Add the wheat germ and mix until just blended. Scrape the smoothie into a glass.

Note: To make pomegranate ice cubes, pour pomegranate juice into plastic ice cube trays (preferably the easy-to-release variety). Freeze until completely firm; then unmold the cubes onto a cold pan or plate. Transfer the cubes to a re-sealable freezer bag and store them in the freezer. Use as needed.

Note: To toast wheat germ, spread ¼ cup (or more) wheat germ in a single layer on a rimmed cookie sheet. Bake in a preheated 350 degrees F oven for about 5 minutes.

index

Orange Juice

Açai, Cantaloupe, Banana, Orange, and Vanilla Yogurt, **66**

Açai, Peach, Mango, Orange, Peach Soy Yogurt, and Protein Powder, **138**

Açai, Strawberry, Mango, Orange, and Mango Sorbet, **55**

Peach

Açai, Apricot, Banana, and Peach, **14**

Açai, Apricot, Banana, Peach, and Peach Yogurt, **57**

Açai, Apricot, Peach, Banana, and Tropical Sorbet, **17**

Açai, Apricot, Pineapple, Peach, and Peach Sorbet, **18**

Açai, Blackberry, Peach, Banana, Tofu, and Vanilla Soymilk, **104**

Acai, Blueberry, Apricot, Peach, Tofu, Vanilla Soymilk, and Vanilla Soy Yogurt, **108**

Açai, Blueberry, Pineapple, Peach, and Peach Sorbet, **29**

Açai, Blueberry, Pineapple, Peach, Pomegranate, and Peach Yogurt, **65**

Açai, Blueberry, Pineapple, Peach, Tofu, and Peach Soy Yogurt, **112**

Açai, Cantaloupe, Peach, Banana, and Peach Yogurt, **68**

Açai, Cherry, Peach, Banana, and Peach Sorbet, **35**

Açai, Flaxseed, Blueberry, Cantaloupe, Peach, Pomegranate, and Blueberry Soy Yogurt, **126**

Açai, Flaxseed, Peach, Pineapple, Kiwi, Vanilla Soymilk, and Peach Soy Yogurt, **129**

Açai, Flaxseed, Peach, Raspberry, Pomegranate, and Raspberry Soy Yogurt, **130**

Açai, Flaxseed, Pineapple, Peach, Banana, and Peach Soy Yogurt, **131**

Açai, Mango, Blackberry, Peach, Pomegranate, and Peach Sorbet, **37**

Açai, Mango, Blueberry, Peach, Pomegranate, and Zesty Lemon Sorbet, **39**

Tropical Sorbet

Vanilla Soymilk

notes